Foundations and Theoretical Perspectives of Distributed Team Cognition

Foundations and Theoretical Perspectives of Distributed Team Cognition

Edited by Michael D. McNeese, Eduardo Salas, and Mica R. Endsley

CRC Press is an imprint of the
Taylor & Francis Group, an **informa** business

First edition published 2021
by CRC Press
6000 Broken Sound Parkway NW, Suite 300, Boca Raton, FL 33487–2742

and by CRC Press
2 Park Square, Milton Park, Abingdon, Oxon, OX14 4RN

CRC Press is an imprint of Taylor & Francis Group, LLC

Library of Congress Cataloging-in-Publication Data
Names: McNeese, Michael, 1954– editor. | Salas, Eduardo, editor. | Endsley, Mica R., editor.
Title: Foundations and theoretical perspectives of distributed team cognition / edited by Michael McNeese, Eduardo Salas, and Mica R. Endsley.
Description: First edition. | Boca Raton : CRC Press, 2020. | Includes bibliographical references and index.
Identifiers: LCCN 2020013282 (print) | LCCN 2020013283 (ebook) | ISBN 9781138625549 (hardback) | ISBN 9780429459795 (ebook)
Subjects: LCSH: Distributed cognition. | Teams in the workplace—Psychological aspects.
Classification: LCC BF311 .F658 2020 (print) | LCC BF311 (ebook) | DDC 658.4/022—dc23
LC record available at https://lccn.loc.gov/2020013282
LC ebook record available at https://lccn.loc.gov/2020013283

ISBN: 978-1-138-62554-9 (hbk)
ISBN: 978-0-429-45979-5 (ebk)

Typeset in Times
by Apex CoVantage, LLC

Dedication

Just over four years ago I was sitting in an uncomfortable chair listening to the infusion pump load chemicals into Judy (my wife) to combat ovarian cancer. It was hard to concentrate on the ideas I was putting together and my mind wandered incessantly. I was on sabbatical for the fall semester from Penn State. Little did I know that I would be sitting in Mount Nittany Medical Center taking care of Judy as she went through chemotherapy. Fortunately, I was able to devote time as a caregiver simultaneous with focusing on academic goals set for myself while on sabbatical. One of those goals was to create an interdisciplinary handbook that would examine a spectrum of contemporary topics within the cross sections of distributed cognition and team cognition as they applied within given contexts of use. The result of that seed concept is this Handbook of Distributed Team Cognition. *I would like to dedicate this handbook to Judy for the strength, courage, character, and positive outlook she showed during her time of confronting cancer and fighting to get healthy. By the grace of God, she is now cancer-free and healthy again. We just had our 40th anniversary of marriage this May of 2020—we celebrate our mutual love and continued commitment to support each other in every way possible. We look at life as an adventure that may present unexpected yet beneficial experiences that make us better humans. I love you.*

Michael D. McNeese

Contents

Preface

As I began work that focused on theoretical understanding of individual and team cognition in the early 1980s, it was often the case that cognition was framed as a phenomenon unto itself, i.e., *inside-the-head* without context, technology, or affective considerations. Much of the research within team cognition progressed within the traditional cognitive psychology theories of understanding (e.g., Hackman & Morris, 1975; Roby, 1968) and was often articulated as group performance or group decision making. This history and perspective still continues with many theories that abound today. The roots of industrial and experimental psychology often co-mingled to produce early notions of teamwork that were responsive to cognitive frameworks. Other alternative perspectives developed as well (situated and distributed cognition) that took into consideration the power of the context, the social impacts within work settings, and the ecological underpinnings of teamwork. Therein, other perspectives formulated that involved cognition, work, and context. As the focus on work prevailed with insights into technologies and computation, many contextual-based theories began to evolve which gave heightened awareness to social theories and the use of technologies to support teamwork (e.g., the development of computer-supported cooperative work (CSCW) and cognitive systems engineering (CSE)).

During the early 1990s I was appointed as the Director of the Collaborative Design Technology Laboratory, Air Force Research Laboratory/Human Effectiveness Directorate, Wright-Patterson Air Force Base, Ohio. I mention this as this is where theoretical yet interdisciplinary ideas about teams, cognition, design, context, and distributed work started to congeal and become a major passion of mine. While I was acutely aware of how technologies could support teamwork and how theories acted to explain what teams produce, many research approaches and publications were still divorced from context, technologies, and computational-information science concepts. The background and interwoven streams of team cognition and distributed cognition fomenting together has wielded new nuances of exploration while continuing to be relevant for theoretical understanding of team phenomena. Theories will continue to provide insights and research production for new research studies, models, and frameworks, and enhance meaning for how technologies are considered as mutual partners in complex work. This book will continue to look at fundamentals, theoretical concepts, and how theory informs perspectives of thinking about distributed team cognition. The chapters provided yield a broad understanding, and may at times not show complete integration—such is the nature of diverse thinking. On the other hand, they may have points of convergence, intersection, and even integration where it can make sense.

As I reflect back upon our own journey, our own early work began with theoretical cognitive psychology approaches in the team decision-making area wherein concepts revolved around situation awareness, cognitive overload, attention, memory, and judgment in contextual-free tasks. As ecological psychology became more important and naturalistic decision making starting to gain momentum, the roles of context and technology became a major consideration for distributed team cognition (e.g., electronically mediated communication for remotely located and distributed teams; learning and knowledge transfer in cooperative learning). However, one early research approach to teamwork (McNeese & Brown, 1986) attempted to accomplish what this current book represents in terms of deriving cogent intersections and integration of perspectives—where it makes sense. Our technical report did a complete review of large group displays and their use in teamwork (from theoretical as well as practical perspectives) within the context of *command and control settings*. It looked at past work where it was relevant, considered current research findings that could be applied, and projected user experiences for future applications of a collaborative technology. This early work set a tone—more than 30 years ago—that continues with this book and the set of books which foments the *Handbook of Distributed Team Cognition*. Note—I am very pleased and fortunate to have Dr. Clifford Brown, Professor, Wittenberg University contribute to this volume as well as I consider him a valued mentor and trusted friend.

Michael D. McNeese
February 2020

SPECIAL NOTIFICATION

This book was produced during the outbreak of the COVID-19 crisis that caused many people around the world by necessity to participate in distributed team cognition in their everyday lives. Many of us were required to be separated by distance to avoid the spread of the virus, hence collaborative work/meetings, joint entertainment, church services, and other activities were conducted through the use of distributed tools, technologies, and apps. As such, the topics within this book are highly relevant for the times we live in and are experiencing. Technologies such as Zoom and Skype facilitated connectedness, teamwork, and social awareness that enabled life to continue in the best possible way. As we adapt to the circumstances of this virus, perhaps many elements of distribute team cognition will be inculcated as part of our permanent culture/society. As we face the summer of 2020, the trajectory of COVID-19 is uncertain and indeterminate. We wish all those affected and impacted by COVID-19 the best path forward.

Editors

Michael D. McNeese is a Professor (Emeritus) and was the Director of the MINDS Group (Multidisciplinary Initiatives in Naturalistic Decision Systems) at the College of Information Sciences and Technology (IST), The Pennsylvania State University, University Park, PA. Dr. McNeese has also been a Professor of Psychology (affiliated) in the Department of Psychology, and a Professor of Education (affiliated) in the Department of Learning Systems and Performance, at Penn State. Previously, he was the Senior Associate Dean for Research, Graduate Studies, and Academic Affairs at the College of IST. Dr. McNeese also served as Department Head and Associate Dean of Research and Graduate Programs in the College, and was part of the original ten founding professors in the College of IST. He has been the principal investigator and managed numerous research projects involving cognitive systems engineering, human factors, human-autonomous interaction, social-cognitive informatics, cognitive psychology, team cognition, user experience, situation awareness, and interactive modeling and simulations for more than 35 years. His research has been funded by diverse sources (NSF, ONR, ARL, ARO, AFRL, NGIA, Lockheed Martin) through a wide variety of program offices and initiatives. Prior to moving to Penn State in 2000, he was a Senior Scientist and Director of Collaborative Design Technology at the USAF Research Laboratory (Wright-Patterson Air Force Base, Ohio). He was one of the principal scientists in the USAF responsible for cognitive systems engineering and team cognition as related to command and control and emergency operations. Dr. McNeese received his Ph.D. in Cognitive Science from Vanderbilt University and an M.A. in Experimental-Cognitive Psychology from the University of Dayton, was a visiting professor at The Ohio State University, Department of Integrated Systems Engineering, and was a Research Associate at the Vanderbilt University Center for Learning Technology. He has over 250 publications in research/application domains including emergency crisis management; fighter pilot performance; pilot-vehicle interaction; battle management command, control, communication operations; cyber and information security; intelligence and image analyst work; geographical intelligence gathering, information fusion, police cognition, natural gas exploitation, emergency medicine; and aviation. His most recent work focuses on the cognitive science perspectives within cyber-security utilizing the interdisciplinary Living Laboratory Framework as articulated in this book.

Eduardo Salas is the Allyn R. and Gladys M. Cline Chair Professor and Chair of the Department of Psychological Sciences at Rice University. His expertise includes assisting organizations, including oil and gas, aviation, law enforcement, and healthcare industries, in how to foster teamwork, design and implement team training strategies, create a safety culture and minimize errors, facilitate learning and training effectiveness, optimize simulation-based training, manage decision making under stress, and develop performance measurement tools.

Dr. Salas has co-authored over 480 journal articles and book chapters and has co-edited 33 books and handbooks as well as authored one book on team training. He is a past president of the Society for Industrial and Organizational Psychology (SIOP) and the Human Factors and Ergonomics Society (HFES), and a fellow of the American Psychological Association (APA), Association for Psychological Science, and HFES. He is also the recipient of the 2012 Society for Human Resource Management Losey Lifetime Achievement Award, the 2012 Joseph E. McGrath Award for Lifetime Achievement for his work on teams and team training, and the 2016 APA Award for Outstanding Lifetime Contributions to Psychology. He received his PhD (1984) in industrial/organizational psychology from Old Dominion University.

Mica R. Endsley is the President of SA Technologies, a cognitive engineering firm specializing in the development of operator interfaces for advanced systems, including the next generation of systems for military, aviation, air traffic control, medicine, and power grid operations. Previously she served as Chief Scientist of the U.S. Air Force in where she was the chief scientific adviser to the Chief of Staff and Secretary of the Air Force, providing assessments on a wide range of scientific and technical issues affecting the Air Force mission. She has also been a visiting associate professor at MIT in the Department of Aeronautics and Astronautics and associate professor of industrial engineering at Texas Tech University. Dr. Endsley is widely published on the topic of situation awareness and decision making in individuals and teams across a wide variety of domains. She received a PhD in industrial and systems engineering from the University of Southern California. She is a past president and fellow of the Human Factors and Ergonomics Society and a fellow of the International Ergonomics Association.

Contributors

Clifford E. Brown
Department of Psychology
Wittenberg University
Springfield, Ohio

Jeffrey R. S. Brownson
Department of Energy and Mineral Engineering
The Pennsylvania State University
University Park, Pennsylvania

Lisa A. Delise
School of Business
Meredith College
Raleigh, North Carolina

Tristin. C. Endsley
Charles Stark Draper Laboratory
Cambridge, Massachusetts

Katherine Hamilton
College of Information Sciences and Technology
The Pennsylvania State University
University Park, Pennsylvania

Caroline A. Hammett
Department of Industrial, Mechanical, and Systems Engineering
University of Rhode Island
Kingston, Rhode Island

Verlin B. Hinsz
Department of Psychology
North Dakota State University
Fargo, North Dakota

Gretchen A. Macht
Department of Industrial, Mechanical, and Systems Engineering
University of Rhode Island
Kingston, Rhode Island

Jacqueline Marhefka
Department of Psychology
The Pennsylvania State University
University Park, Pennsylvania

Michael D. McNeese
College of Information Sciences and Technology
The Pennsylvania State University
University Park, Pennsylvania

Nathaniel J. McNeese
College of Computing
Clemson University
Clemson, South Carolina

Susan Mohammed
Department of Psychology
The Pennsylvania State University
University Park, Pennsylvania

John Teofil Paul Nosek
Department of Computer and Information Sciences
Temple University
Philadelphia, Pennsylvania

S. Orestis Palermos
School of English, Communication, and Philosophy
Cardiff University
Cardiff, Wales, United Kingdom

James A. Reep
College of Information Sciences and Technology
The Pennsylvania State University
University Park, Pennsylvania

Joan R. Rentsch
School of Communication Studies
University of Tennessee
Knoxville, Tennessee

Rachel Tesler
Department of Defense
Rockville, Maryland

Dominique Engome Tchupo
Department of Industrial, Mechanical, and Systems Engineering
University of Rhode Island
Kingston, Rhode Island

R. Scott Tindale
Department of Psychology
Loyola University Chicago
Chicago, Illinois

Jeremy R. Winget
Department of Psychology
Loyola University Chicago
Chicago, Illinois

Primer (Introduction)

To provide a quick introduction and basic overview of each of the three volumes that constitute this handbook, we utilize a *primer.* The primer contains our concept for each book and then gives short summaries of the chapters within the book. This provides a handy way to get a sense of what each book will contain in terms of topics, together with the focus and authors of each chapter.

For volume 1, *Foundations and Theoretical Perspectives of Distributed Team Cognition*, the purpose is to begin the handbook with foundational knowledge that is relevant to understanding distributed team cognition. Inherently, this encompasses history, theoretical viewpoints that present multiple perspectives, basic research knowledge and findings, and unique ideas or philosophies that underlie the development of distributed team cognition. Volume 1 presents a lot of background information, definitional reviews, and historical threads, with an intent towards conceptualization of what we mean by distributed team cognition and how differing approaches toward it can allow the reader new ways to think about it. In this sense it is intended to generate the new out of the old and afford integrative bridges across time. By doing that the reader can experience what led to the development of this handbook and the focus of the topics within while at the same time be shown where it is headed into the future. Volumes 2 and 3 then help to color in more specifics on current research topics and applications.

As the first editor of the handbook, McNeese devotes Chapter 1 to the specifics underlying the development or emergence of *distributed team cognition* as the composite integral of distributed cognition and team cognition, and lays down the beginning girders upon which the formative structure of the handbook is supported. He utilizes his own autobiographical history and longitudinal development in design, cognition, and collaborative technologies to illustrate how one specific type of multiple perspective is formulated. Many of the basics that underlie topics throughout the book are addressed in this first chapter.

The second and third chapters by McNeese, McNeese, Delise, Rentsch, and Brown are heavily steeped in the history and traditions of both team cognition and distributed cognition through the lens and practice of team simulations. Chapter 2 focuses on team simulation and the first authors experiences at Wright-Patterson Air Force Base wherein teams, technologies, and context emerged as a distinct research entity, and where much of the early history of this area was conducted. Chapter 3 provides a cogent overview of distributed team cognition as a research focus within team simulation and gives examples of real-world simulation while and surveying major research variables studied. Chapter 3 also outlines a research framework that can be used for interdisciplinary study of distributed team cognition.

The fourth chapter by Tindale, Winget, and Hinsz provides a deep dive into understanding what teamwork means from the perspective of team information processing, highlighting how different tasks and the amount of tasks engaged in influence team performance. This chapter is very useful in that it touches on a lot of the concepts inherent in distributed team cognition and the patterns underlying the relationships among motivation, task demands, shared knowledge, coordination, and specific judgments in decision making or problem solving.

The fifth chapter by Palermos is unique in that it entwines the philosophy of mind, the group mind, functionalism, and extended cognition and posits alternative ideas to distributed cognition for healthy counterpoints. An example utilizes the bee colony as a point of reference for how they make decisions on finding their next nest. One of the qualities of the handbook this chapter exemplifies is presenting alternative viewpoints that signify worldviews one can utilize when thinking about distributed team cognition.

The sixth chapter by Nosek is one of the viewpoints represented in the handbook that slants towards ecological interpretations of group activity and highlights Collaborative Action Theory in the social construction of knowledge. The chapter applies this theory and highlights the use of affordances as related to user experience, design, and information overload hence again providing alternative suggestions to a strict cognitive paradigm.

The seventh chapter by Marhefka, Mohammed, Hamilton, Tesler, Mancuso, and McNeese returns to a more cognitive-based philosophy and looks at the impacts of perceiving and sharing team mental models within an interrelated team task (the NeoCITIES simulation) wherein temporality could be examined and assessed. The chapter provides a compelling example of distributed team cognition experimentation from hypothesis formation to setting up a team experiment using a simulation to measurement to data analysis. Temporality is very important for understanding distributed team cognition and this highlights findings in the chapter.

The eighth chapter by Reep provides yet another facet of teamwork and distributed cognition, the role of expertise. The chapter looks at how expertise develops from novices to experts and what this means for knowledge organization, pattern recognition, reasoning, and other abilities that come into play when a team needs to integrate individuals with varying capabilities and experiences. A future research agenda is provided to enhance continued study in this area as related to distributed team cognition.

The ninth chapter by Endsley, Macht, Engome Tchupo, Hammet, and Brownson continues with the thought that teams consist of individuals that have to integrate their backgrounds, knowledge, skills, abilities, and cultural differences in an effective way to address the complexity of work. This is often not easy and can challenge a team to complete tasks so this is an important foundational aspect of how cognition is distributed and shared. Research involving diversity in teamwork revolves around a focus on race, ethnicity, culture, cognitive styles, and personality. The chapter looks at how these variables can influence and determine the outcome of a team's performance.

This concludes the primer for Volume 1. Many of the principles and ways of construing distributed team performance are articulated in this first book hopefully evoking inquisitive thinking while at the same time generating learning about what this handbook intends to communicate.

Michael D. McNeese

1 Distributed Team Cognition
Integration, Evolution, and Insight

Michael D. McNeese

CONTENTS

INTRODUCTION

In 2001, Michael D. McNeese, Eduardo Salas, and Mica R. Endsley published a book entitled *New Trends in Cooperative Activities: System Dynamics in Complex Environments*. That book was designed to capture state-of-the-art research involving team cognition as it occurred in dynamic environments. Nineteen years have passed since publication of that volume and a lot of things in the world have changed dramatically. The passage of time and the presence of disruptions provides new zeitgeists for understanding how teamwork and team cognition have evolved given:

- consideration of theoretical frameworks,
- derivation and reification of cognitive science concepts/cognitive processes,
- methodological diversification and innovation,

- progression of knowledge capture tools and measurement of different phenomena,
- socio-cultural embodiment in various fields of practice, and
- the rapid design/development of information, communication, and cognitive technologies.

These evolutionary imprints jointly contribute to what is meant by distributed team cognition. As we embark on this current journey 19 years later to collectively review and comprehend distributed team cognition, it is imperative to be mindful of the history and roots from whence it came. Therein, while the authors of the chapters within necessarily point toward the contemporary mindset they also flow from the foundation and work of many scientists who came before them. History is important and should be recognized and not forgotten. At the same time, it is the pull of the future that challenges us to make progress within this exciting area of science and research.

The *Handbook of Distributed Team Cognition* provides a refreshing look at what has transpired within team research as informed through cognitive activity and technological advancement, as distributed across the digital global society. Our intent is to enlighten the reader with the specifics of evolutionary thought that demonstrate how team cognition has been transformational in its impact on human beings. The purpose of this first chapter is to look at some of the philosophical underpinnings that led to producing this handbook. Philosophical underpinnings necessarily are generated from the worldviews that an author believes in, the history and development of work, and how important disruptions cause individuals and society to change how they interpret and defend reality. To demonstrate this idea, I will provide a personal-developmental look at how my own passions and experience eventually evolved into research within the distributed team cognition genre. In addition to philosophical underpinnings, this chapter builds a foundational footing and flooring of the handbook in terms of what to expect. We provide rationale for why this undertaking was initiated and needed. It is our hope that the handbook induces new thinking and inspiration to further cognitive science while reviewing cogent research in terms of theory, methods, measurements, and applications.

BEGINNINGS

The focus of research and development related to distributed team cognition derives from the attention and keen awareness that any given researcher takes longitudinally across time, thought, and place. And this may place each of us in a distinct realm of meaning for a given topic, i.e., our understanding of a phenomena, process, or technology but based on scientific evidence, not opinion. The color of meaning derived absolutely depends on many influences, disruptions, adaptation and assimilation of new directions with supportive research results, practical considerations, and the value accorded to many biases, streams of thought, beliefs, and experiences that together formulate what is referred to as knowledge. While knowledge exists at the general, principled level across many studies and replications, it exists at specific and personal levels as well. Specific knowledge helps human beings (and computational) agents reason about the world around them (their sense surround) in ways that coincide with their constructed interpretation. Personal knowledge is construed and constructed according to the contexts we experience and derive—over and over again. It is specific and subjective but also can be generalized to the point of objective truth in ways that make sense to many, and again is distilled across time and culture. This distillation can change the outcome of the construction over time so what was true 20 years ago may not hold today. Specific and personal knowledge is therefore malleable in formation and changes based on what you already know (and resides in the memory that contains it). Having said this, knowledge is sharable across cultures, groups, and teams accordingly. In fact, many of our concepts for this handbook derive from basic level ideas surrounding shared knowledge among agents (e.g., common ground, team mental model, team schema, common operational picture, jointly constructed memories, and so on). What agents do with the knowledge they have or share dovetails into action, achievement, and performance.

Knowledge as design (Perkins, 1986) takes what we know (individual and team level) to the state of what may be accomplished (activity and performance) through what can be produced (designs), and in turn how well a design may be leveraged by actors in a situated context (use). This premise is directly coupled to ideas resonate within the research areas of distributed cognition (Hollan, Hutchins, & Kirsh, 2000), team cognition (McNeese, Salas, & Endsley, 2001; Salas, Fiore, & Letsky, 2013), and interactive team cognition (Cooke, Gorman, Myers, & Duran, 2013). Ideas about distributed cognition and team cognition appear throughout this handbook in the form of what we know, how we generate designs for use, and how we act in a wide variety of applications and fields of practice. Distributed cognition involves the individual as related to the broader sense-surround wherein an individual is embedded within a given environment. That often includes the necessity that workers leverage individual work with others (i.e. working together) to achieve a desired outcome (goal) where complex and complicated interdependencies are operative. And the outcome is resonant with using cognitive processes to assess situations, make decisions, or solve problems. This larger world of interaction frequently revolves around the dynamics of teamwork and reveals collective, cultural, and socio-technical values if it rings true (i.e., reflects universality).

Philosophical Quest of the Handbook

In some ways, a handbook could be construed as generating a five-dimensional maze where time = theories/perspective, width = approach/methods, length = applications, and height = technologies. If you can imagine a 3-D block that changes dimensions through time you can induct the construct of a matrix that emerges according to given research foci. But to travel through the matrix one might choose any single point within the block to represents an instantiation point. If one imagines the center the block as a place to start (with a given theory, method, application, and technology utilized), as you move in any direction you change one of these dimensions and therein alter the research focus taken. As one moves across time with differing dimensions of these constructs being utilized for their own research niche, the fifth dimension—from instantiation-to-abstraction—can be hypothesized for a researcher. One might look at this for a specific researcher but in the concourse of a handbook this matrix represents the population of many worldviews that have been derived across set timelines, hence an abstraction of abstractions. Obviously, this only represents a select subset and not all possible viewpoints. Therein, the fifth dimension indicates an abstraction or generalization emanating from a given point within the overall matrix to a collectivist derivation of topics that are culturally current and relevant to researchers who share interests in distributed team cognition.

Stated another way the matrix shows the evolution of science across time from a concrete point to levels of abstraction. That is, at any given point a reader is surrounded by contiguous spectra that are highly related to a researcher's specific interests—in this case distributed team cognition concepts. The further out from a given point—in all dimensions—suggests a greater distance or abstraction from the specifics that a given researcher has taken and is engaged in. As one takes in spectra positions there may be segues for generalization or even general knowledge can be transmitted across time and demonstrate the qualities of invariance or veridicality. One may enter the matrix through different portals that transport one to different sectors representative of a select spectra of research. Sectors can be contiguous such that passages interconnect sectors to traverse complex interdisciplinary research space.

Therein, the handbook taken as a whole does represent a complex interdisciplinary research space and contains all the elements above with the intent to inform the reader at the general knowledge level as well as the specific research spectra level. Hopefully it provides the reader with a thoughtful circulation of ideas, conceptual trajectories, and boundary constraints evident in distributed team cognition. The 5-D matrix also functions as a mirror in that it allows a reader to assimilate the content they read with what they already know for themselves with respect to a given niche (the process associated with expanding their own internal concept map for a research area or topic). This affords generative learning, reflection, and metacognition as one compares, processes, and integrates differing points of view within the handbook with one's own centric point of view (i.e., beliefs) to determine the degree of fit. Reflection is certainly dependent upon each of our backgrounds, abilities, training, experiences, culture, and biases and therein anyone's assimilation of the *Handbook of Distributed Team Cognition* can be different but also contain common ground, just as our individualistic cognition is different, and yet at some degree of abstraction is the same. (We enter the handbook via portals closest to our way of thinking most likely.)

Specific openings represent certain intellectual spaces that are relevant for a given aspect of the handbook but do not necessarily generalize and therein do not represent the holistic overall intellectual space. Because each of us represent a unique set of individual differences—the ways we think about distributed team cognition, the manner by which we go about conducting research, and the concepts we produce for technological designs to support distributed cognitive activities can necessarily be different from other researchers. At the same time as research scientists, professional educators, engineers and designers, and futurists we share a common ground of knowledge that sets foundations that are meaningful in the broader sense. We hope to capture multiple perspectives from many viewpoints as well as formulate a contemporary common ground of knowledge for the reader, within the books that comprise the handbook. For you the reader, the handbook's purpose is to provide multiple, useful examples of research that generate principles, processes, and practical applications within the distributed team cognition research space.

Our own guidance in understanding this complex interdisciplinary research space came from advisors, mentors, and colleagues who watched over us and pointed us in the right direction. It is our hope that the authors presented in this handbook add to your level of knowledge and therein provide additional guidance to expand understanding. What we thought we knew many years ago certainly has evolved over time, being shaped and refined through these influences, relationships, consequences, and the truth of our lives. As one reads through the chapters resident throughout the three volumes that comprise the handbook, the prose traces through the specific heritage of learning, perspective, and practice for a set of authors. Because the scope of the handbook addresses and drills down on interdisciplinary research topics, the academic disciplines represented span across cognitive science, systems engineering, psychology, computer science, information-communication technology, human-computer interaction, anthropology, human factors, and design, in addition to other cogent disciplines as needed.

As we contemplated the focus and direction for the handbook one of the central principles considered was to supply multiple perspectives and yet show precipitous integration and even glances of invariance within a topic at hand. It is our hope that the multiple perspectives provided herein can prosper the reader, activate the curious minded, and at the same time cover a large expanse of intellectual space within the distributed team cognition genre while making sense of it all. As you read through the handbook it is akin to a labyrinth where passages connect to synthesize integrative thoughts that ring true.

Direction and Structure

The previous section provided background, motivation, inspiration, and some of the historical layers underlying the development of the handbook. The next few sections focus more on foundations, content, rationale, and the emphasis taken in the overall composition. While information in these next sections may overlap it is intended to show the reader the why, what, who, and how of distributed team cognition. To begin we have organized the overall handbook to consist of three individual volumes. One can obtain all of the books or just the ones that generate interest. One will find some topics to be more distributed across the three volumes than others (after all it is distributed cognition!). We have structured the three volumes in the *Handbook of Distributed Team Cognition* topically as follows: Volume 1, Foundations and Theoretical Perspectives; Volume 2, Contemporary Research—Studies, Methods, Models, and Measures; and Volume 3, Fields of Practice and Applied Solutions. These three volumes span relevant, current topics in the information continuum within distributed team cognition. At the beginning of each volume a primer is provided which outlines the concept of the volume and provides a brief synopsis of each chapter.

What Is the Focus of the Handbook?

The primary direction undertaken within this handbook focuses on the interdisciplinary study of distributed team cognition as defined by the interrelationships and appropriate action potentials inherent in information, technology, people, work, and context. The handbook approaches and defines current worldviews of distributed cognition as they are related to what cognition is, what cognition does, and how it is distributed and carried out as part of teamwork. As such the definition of distributed cognition is cognitive activity (individual, teams, computational agents) intended to address or overcome complex challenges, problems, or constraints that are distributed across and influenced by a specific context or environment. Distributed team cognition is used in many different application domains, but one example of this would be the way police work to understand a crime, the crime scene, the perpetrator, how the analytics of the situation emerges especially across different venues and with different timelines, and how distributed cognition and specific information and communication technologies play out and are used in conjunction with distributed cognition explanations. This is no more evident than with the real-world example of the Boston Marathon bombings which took place in 2013 whereupon multiple teams worked together to gain an understanding of a continuously emergent situation that spanned across multiple contexts and required the use of a variety of technologies (e.g., face recognition technology) to make sense of the analytics underlying decision making, problem solving, context, and judgment.

One of the strong components within the handbook is the role of technology as a disruptive agent that transforms team cognition in terms of definition, process, and application. Technology has been distributed across our lives in so many ways and is partly responsible for the revolution in the digital global economy. Its use makes our lives easier and more complicated all at the same time. We may take it for granted and not realize when or how it has been important to address many real-world problems encountered at the societal and individual level. Technology does not exist in a vacuum but has been innovated as a means to an end. It can and has worked for both good and evil. Unfortunately, all too often it has not been designed with much consideration for human needs or team capabilities but simply evolved just because it was possible to do so. While this may be good for purposes of invention, the human use of technologies has historically not been a prime consideration although this has changed in the last half century owing to the practice of human factors engineering (Meister, 1999; Wickens, Hollands, Banbury, & Parasuraman, 2013).

This handbook provides the landscape for how technology has been positively influenced by information, people, and context in many fascinating and productive ways. In particular, it examines how (1) teams have put technology to use with focus on distributed applications and (2) how technology has amplified the capabilities for quick and efficient distributed information within problem solving and decision making. It focuses on cognitive processes that teams use as they are distributed across contexts, tools, and apps, and recognizes the varieties of information representation that are now possible to enhance problem comprehension. Taken together the handbook weaves a tapestry of interdisciplinary research which we refer to as distributed team cognition to capture these notions and more. While team cognition and distributed cognition have their own distinct entities (our research groups have published within each genre respectively) current research is now reflective of the integrative coupling of these areas (more on that later).

What Do We Mean by Distributed Cognition?

Distributed cognition is relative to human, social, cultural, and organizational factors that dynamically emerge within the context and are often contingent upon availability of data, information, and knowledge resident within a technological milieu. Cognition is distributed across many planes and subject to dynamic fluctuation given constraints that are active at any given moment. Distributed cognition is not about cognition only viewed as resident within the brain but takes the broader view that cognition is embodied (see Varela, Thompson, & Rosch, 1991; Wilson, 2002), represented within the situated context where it is operative, and is entwined with activity. While there are many other definitions cogent to distributed cognition (Hutchins, 1995; Zhang & Norman, 1994) the definition stated in the previous section emphasizes the mutual interplay among cognition, culture, and artifacts with the understanding that activity develops through dynamic interaction in the midst of multiple constraints. This particular definition is similar to ideas put forth under the auspices of ecological psychology, situated cognition, enactive cognition, and situated learning. These are the footers that formulate my own worldview and in turn have led to the motivation to produce a handbook that has broad precepts. This includes giving credence to other worldviews as well. Inherently, the philosophy and viewpoints of any chapter herein may vary according to influences and history of the individuals composing the chapter. The intention is to capture diverse viewpoints rather than promulgate one specific framework. While this is a goal one can also derive levels of integration across the chapters that are relevant for certain concepts, principles, processes, methods, measures, application, and practice even though different contexts that are utilized reflect unique histories and findings.

Many worldviews (i.e. philosophies, conceptualization, and integrated system of beliefs) have been used to derive distinct connotations for questions such as (1) What is cognition? (2) Where does cognition exist? (3) What does cognition produce? (4) How is cognition constructed? (5) How is cognition carried out in the real world and distributed as part of teamwork (distributed team cognition)? While the first four meta-level questions are extremely important and actually presuppose question five, it is question five that really generates the foundation and formulates the focus of this handbook. Indeed our own work suggests that cognition is strongly coupled with cooperative activities (team cognition, McNeese et al., 2001; Salas, Cooke, & Rosen, 2008), distributed and coordinated across multiple vantage points according to need (distributed cognition, Hollan et al., 2000), situated within specific contexts (situated cognition, Brown, Collins, & Duguid, 1989), built through continuous experience as shaped by individuality, history, learning, and culture (Lave & Wenger, 1991), predicated on the availability of data, information, and knowledge (information sciences, Zins, 2007), and is extended or augmented through the use of technology (information and communication technologies, mobile devices, ubiquitous computing, computational agents, the worldwide web, social networks, etc.).

Yet there are few volumes that provide multiple perspectives across various worldviews in an attempt to understand the complex interrelationships that are distributed across cognition, information, technology, people, work, and context. The handbook incorporates trends, topics, and trajectories that span across this broad panorama and includes topics from information and cognitive sciences, industrial-organizational psychology, computer science, computer-supported cooperative work, human-computer interaction, learning science and gaming, and other cogent viewpoints.

The topics that resonate in the handbook provide an interdisciplinary nexus that articulates contemporary directions in distributed team cognition as relevant for the society we live in. The handbook examines and reviews unique worldviews of cognition (the why), presents unique state-of-the-art theories, research, and technological innovations that inform distributed cognition (the what), highlights innovative methodologies in use (the how), and provides examples of applications and cases within specific fields of practice (the use) as understood by recognized world-class experts (the who). Together the topics—as reflected within a given chapter—formulate collective frame of minds that represents a new synthesis of cognition.

What Are the Underlying Needs for a Handbook?

The primary emphasis of the first editor's scholarship over the last two decades has examined teams, work, and cognition through particular research lenses: situated cognition (Brown et al., 1989; Young & McNeese, 1995), team cognition (Salas, Fiore, & Letsky, 2013; Mohammed, Hamilton, Tesler, Mancuso, & McNeese, 2015), situation awareness (Endsley, 1995; McNeese & Perusich, 2000), and ecological basis of cognition (Hutchins, 1995; McNeese & Forster, 2017). The research underlying these perspectives has advanced theoretical positions that emphasize the roles that social factors, information sciences, experience, and technology have in the development and support of teamwork such as problem solving (Young & McNeese, 1995), decision making (McNeese, Mancuso, McNeese, Endsley, & Forster, 2014), or the transfer of knowledge (McNeese, 2000). Historically, this work is traceable to the intellectual pursuits of John Dewey (1938) and the interdisciplinary problem-based learning theories of John Bransford (Bransford, Brown, & Cocking, 2000).

A strong part of advancing the socio-cognitive and socio-technical aspects of teamwork necessarily includes related focus on integrative methodologies that provide ecological and concomitant interpretive powers (e.g., see AKADAM, advanced knowledge acquisition and design methodology, McNeese et al., 1995; A Framework for Cognitive Fieldwork, McNeese, Bautsch, & Narayanan, 1999; The Living Lab Approach, McNeese, Mancuso, McNeese, Endsley, & Forster, 2013). Other components of this focus transact across social psychology and how it influences intelligent systems (Wellens & McNeese, 1987, 1999; McNeese, 2006). Yet the overwhelming intent of this work is to expand interdisciplinary connections to make distributed team cognition more meaningful to our everyday lives, and to enhance and understand more clearly the roles of context, socially coupled systems, and information and communications technologies (e.g., Fan, McNeese, & Yen, 2010).

Our own work includes many studies, dissertations, and research grants appropriated for an integrated framework to examine distributed team cognition that emphasizes (a) cognitive systems, (b) contextual fields of practices, (c) socio-technical/socio-computational studies, and (d) human-in-the-loop simulations and modeling. To further extend and highlight the work mentioned above my co-editors provide an even more extensive bandwidth with much recognition and highly cited studies in the areas of team cognition, training, situation awareness, and simulation studies that complement, reify, and provide sound coverage of team cognition work over the last 25 years. That work for all of us has spanned across multiple positions in government, business-industry, and academia so it is inclusive of many types of experiences, layered understanding, and positions on the topic. Indeed, our joint book in the early part of the 2000s (McNeese et al., 2001) provided foundations and trends within an emergent team cognition community. Almost 20 years later it is important to assess what has evolved, what is important now as the state of the art, what may come to pass as the future of distributed team cognition, and how team cognition has inherently become more distributed with the influence of information sciences and technology.

We have produced many papers, given many presentations, and received grants that clarify various directions mentioned above. Yet rarely are the multiple perspectives and worldviews that underlie distributed team cognition available for consumption within a single volume. Therein the *raison d'etre* for this handbook can be delineated (i.e., "Why this book now?"). The first justification is based on the idea of thoughtfully designing a handbook that brings together top scholars to present multiple perspectives on team cognition in one fell swoop (what was referred to earlier as collective frame of minds). This alone would provide a useful and current research volume to specify the logical connections among worldviews, theory, methods, application, and practice.

As we consider recent work from all of us in areas of distributed team cognition, team mental models, team situation awareness, information sharing, and decision making, it is only natural to think about how significant the reach of team cognition has been and to examine how it is extended through the influence and application of information sciences and information technologies. Therein, the second justification for the handbook is to show how information sciences and technology has evolved team cognition to be more in line with distributed cognition areas. Breaking this down in perhaps a deeper way we can now say that "distributed" connotes more than just cognition distributed across place (i.e., the contextual and cultural basis for distributed cognition, see Scribner & Tobach, 1997). It is true that one of the primal meanings of distributed cognition is that of knowing how thoughts are linked to work in context. But being distributed also means more than just that. It means that cognition is distributed across time and place (temporality proceeds into the future but is often reliant on the past, often referred to as temporal work, see Mohammed et al., 2015). Cognition necessarily distributes information where collaborative information must be sought, defined, retrieved, reified, foraged, and saved (often referred to as collaborative information seeking, see N. J. McNeese & Reddy, 2017, and information fusion, see Hall & Jordon, 2010) and information is distributed across an array of team members (participants interacting towards a joint goal where common ground emerges along the way). Cognition is distributed from an individual to a team and back to individuals in varying degrees of interdependency (often referenced as knowledge transfer and collective induction, see McNeese, 2000). And finally, cognition in today's society is distributed across many levels of computation and technology (e.g., social informatics, web and mobile computing, artificial intelligence, the rise of enterprise networks, big data analytics, and the internet of things all can produce new ideas and conceptualizations of what cognition is and what it means as well as how group–team definition varies) and is reified to new levels through dynamic collaborative interaction.

Looking at this point with a bit more depth, distributed technologies such as the internet have fundamentally changed what we mean by team cognition and how it is practiced in 2020. Technologies have been disruptive towards defining what cognition means in 2020. First, distributed cognition is now very much influenced by virtual work that transpires through the providence of ubiquitous computing, social informatics, the internet of things, and smartphone applications. Unfortunately, many of the original theoretical perspectives on team and distributed cognition that emanated out of psychology and communications communities are entrenched in the assumptions of the computational substrate of the 1980s, and hence are rather uninformed by monumental gains in virtual interaction, data science, robotic enablement, crowdsourcing, and human-autonomous systems.

Second, a lot of work in distributed cognition is from unary points of view that tend to not be interdisciplinary but often are subject to certain biases within a given scientific community. For example, the communities of computer-human interaction (CHI), computer-supported cooperative work (CSCW), human factors (HF), situated cognition, cognitive science and modeling (CogSci), cognitive systems engineering (CSE), and naturalistic decision making (NDM) all touch upon distributed cognition but have different mindsets, methods, biases, and levels of understanding and are often insular to each other rather than facilitating mutual learning. Their funding profiles and sources may also highlight directions and outcomes that are not similar. This can result in guarded values and closed minds that are typically not open to discussion, discernment, and hermeneutics that could advance further growth and development across worldviews.

Third, "cognition" is not typically seen in very broad terms but frequently is only equated to traditional cognitive studies that are grounded in memory, learning, reasoning, and judgment. While these areas are genuinely important and have been researched extensively, it is imperative that the other powers of the mind are examined, such as creativity, intuition, values, imagination, emotion, and culture, wherein collective mindfulness is a useful construct.

Fourth, the areas of application for distributed cognition concepts are very wide open and expanding with the spread of technological innovation. For example, one of our groups at Penn State embarked on a unique twist of this kind of work: cyber situation awareness (see McNeese and Hall, 2017), as have others but it is just beginning to scratch the surface of interdisciplinary applications. Other application areas/domains that are emerging for consideration of distributed cognition theory and concepts are intelligent systems in natural gas that interconnect data, humans, objects, locations, and supply chains; risk/crisis informatics; health, disease, and medical informatics; climate change and geospatial information science; online interactive learning environments; understanding terrorist social networks; uninhabited vehicles; and human-robot interaction, to name a few. Let's examine some specific contexts in which distributed team cognition could occur.

Examples of What We Mean by Distributed Team Cognition

In addition to the example provided earlier we provide two specific examples of how distributed team cognition is evident within society today, specifically reinforcing the ideas that distributed team cognition is often highly interdependent upon various forms of information and communication technologies. The first example comes from the emergency medicine/hospital domain. An ambulance is called to the scene of a car accident wherein one car has T-boned another. There are multiple injuries that have occurred as each car has multiple occupants. Indeed, this represents an emergency situation wherein time is essential, there are a number of uncertainties with respect to the severity of the injuries, a team of teams approach is necessary (inclusive of initial responders, ambulance crew, help flight crew, and police), an emergent context changes what is known when, and multiple layers of information are distributed across the context, the people involved, the vehicles themselves, and potentially in camera technologies that have recorded the accident. Not only are medical concerns utmost in priority but potential litigious issues will be at play (who caused the accident). This particular situation represents a distributed cognitive event that is dependent on team performance and individual roles within a team, team-to-team transfer of information, and joint problem solving/naturalistic decision making, as well as the use of technology to communicate the state of the emergency across all parties, but also information technology that is used to ascertain situation awareness about the conditions of the occupants. One occupant (a driver) is severely injured and is taken by air-flight to a trauma unit twenty miles from the scene while three other occupants are taken by ambulance to a local hospital. Both the air flight and the ambulance are setup to transmit and convey information (both auditory, visual, and actual photographic imagery) directly to the appropriate hospital units to prepare physicians, specialists, and nurses prior to arrival at the hospital. This is very important for problem solving and decision making particularly with respect to temporal expedience. Therein, this kind of situation is one example of distributed team cognition in an applied domain.

Another contemporary example surrounds the distributed nature of information and cyber security. Multiple hacking incidents have occurred at the highest level of the Department of Defense and other government agencies (on a daily basis). While there are many tools for determining the sources and patterns of hacking and perpetrated attacks, a more nuanced approach would include looking at the human-centered aspects of information security, cyberwarfare, and risk. Some of the work we have partaken (see Cooke & McNeese, 2013; McNeese & Hall, 2017) indicates that these integrated topics of research must indeed be approached from a wholistic perspective representative of distributed team cognition that provides analysis and prognosis to look at (1) interdependent layers of contextual perturbations (inclusive of data, information, knowledge, wisdom), (2) the role of team cognition in both producing and recognizing articulated attacks (strategic offensive and defensive information warfare), and (3) increasing use of a multitude of data science techniques and requisite information technologies that compose distributed cognition to formulate the power underlying situation awareness, risk, and deceptive behaviors. This domain is often only considered from a blunt individualistic, techno-centric methodology that completely ignores the human-centered and distributed cognition approaches needed to break down intent and complex behaviors. The domain has a necessity for considering a team of team's approach to make sense of complex situations, requires a high degree of information sharing for success, and very much requires that a deep understanding of the emergent—often hidden—context that pervades this research area.

Transitional Trajectories

As I began to conceptualize the model for this handbook one thing that struck me is how in the world did I get from early childhood interests and thinking—way back when—to my current level of expertise now? And how did this collective experience come to develop the area for which this handbook has been designed? The authors contributing to the handbook can all ask this same question and the developmental trajectories will obviously be different, but meaningful in terms of the shaping patterns, awareness, and insights that led to research within distributed team cognition. This is how research interests come to pass. To make this extant, the next part of this beginning chapter looks at the longitudinal evolution of distributed team cognition from my own personal, autobiographical outlook. Walker (2017) discusses why autobiography is a legitimate methodological technique to reveal critical research as a narrative inquiry. Walker suggests that "stories reflect a set of values, rules, norms that oversee a person's learning and sense of logic (Maynes, Pierce, & Laslett, 2008)" and "when viewed as source of data, autobiographical narratives situate reflexivity within contexts of cultural settings (DeGloma, 2010) that offer researchers an important set of social and individualized contexts to study (Brockmeier, 2012)" (Walker, 2017, p. 1896). The following critical inquiry seeks to trace historical markers, interdisciplinary connections, and firsthand personal knowledge as pertinent in deriving and constructing this handbook. Necessarily, this example shows how one could come to develop a research niche in distributed team cognition over a long developmental timeframe. Walker emphasizes the point that inquiries such as this provide a basis for identity and lifelong learning. It is inferred then that these components are pivotal in establishing a researcher's awareness, attention to detail, and progression towards expertise within a scientific subject matter.

DEVELOPMENTAL HISTORY: A PERSONAL PERSPECTIVE

As I consider where my life has taken me it certainly is the case that it has been influenced by research and design (R&D) opportunities in many different facets, through multiple distinct venues, and with unexpected outcomes across time (literally over 60 years of shaping influences). As it turns out design has surrounded me in somewhat unusual circumstances prior to the research focus taking hold. Therein, perhaps design and research (D&R) is a better descriptor of a major persuasion on my life. My earliest interests in design occurred through an awareness in imagining inventions and then sketching or drawing them out (early drafting proclivity!). While in grade school I should have paid more attention to teachers but instead—beginning in the third grade—I would lay my head down on my desk and draw and draw and draw. I know this sounds strange but laying my head down enabled me to "see" the design come to life (maybe just a form of concentration). Obviously, this did not help me with assignments and quizzes but the desires of my heart—inventing and drawing—were assuaged.

With Design in Mind

One primary venue for drawing and "first designs" was through sketching and refining cars for many purposes. Cars and car design were my passion and at five years old I had a keen interest in *Car Craft* and *Hot Rod* magazines. They provided motivation for thinking about new designs. As I began reading about performance in cars I started to get engaged with engineering that resulted in improving performance. I also realized that design and performance in cars was intimately coupled with the driver (the human element). Therein, this was the first sense of the human factor as it involved automotive design. In these early days then there was a nexus of form, function, and use. This enabled designs to be transformed into products that accomplished their performance objectives. This was my first foray into user-centered design (building cars around the driver and what s/he was capable of doing). While the engineering design interests continued to unfold I also had creative passion for the aesthetics of car design which correlated more with the field of industrial design. These interests formulated strong seeds that grew into diverse pursuits. I ended up obtaining early degrees in design and technology and psychology, and worked as an engineering designer for several years before the research focus emerged. The seeds of design however took some interesting twists and turns. One of the elements within design that touched my curiosity was the idea that designs could be transformative, both at the individual and the societal level. At the time, I did not realize this consciously but the seeds that set the groundwork for my research expression and evolution where intimately tied to understanding the crossroads of previously distinct vocations or professions, which in contemporary circles today is expressed as interdisciplinary sciences.

I had the fortunate opportunity of growing up in the Dayton-Kettering, Ohio, area which is historically known as a hotbed of design, innovation, and industrial manufacturing (see Bernstein, 1996). Indeed, the city of Kettering was named for none other than Charles Kettering who designed the first self-starter for automobiles and established Delco (Dayton Engineering Laboratories Company) which was the research and design arm of General Motors Corporation. Dayton was the second largest GM town in the country during the 1960s and '70s and was heavily involved in the auto engineering industry. My father worked for Chrysler Air Temp in Dayton, ironically, in the job position of time study engineering.

As I approached ten years old, I developed an awareness of what Dayton was known for in the areas of invention, engineering, and technological progress. The Wright brothers, inventors of the airplane and early flight technology, were pioneers in the innovative spirit behind design. The influence of the Wright brothers was pervasive across many areas of engineering and aviation, and is something native Daytonians are proud of. Growing up in the Dayton-Kettering area as a grade schooler meant that our house was about eleven miles from Wright-Patterson Air Force Base. I started getting interested in flight oddly enough by watching the flight paths of F-4s literally fly over our house as they made their approach to the base. Certainly, there were other aircraft that would fly over as well. But one of the amazing things the F-4s would do is fly faster than sound, hence producing sonic booms. We heard quite a few of these which added to my fascination and intrigue with flight. Another component of this interest was the zeal I experienced visiting the U.S. Air Force Museum (at the time in Fairborn, Ohio). Their displays of actual aircraft describing the history, performance, and precedence of many different flying machines was both incredible and fascinating. This also inspired me to think about new kinds of fast aircraft and make model versions therein. (In addition to creating and drawing designs I started an active hobby of making models—cars, planes, and ships.) The museum also amplified another upcoming interest—space flight—that I found intriguing but was a bit out there when I was ten years old. The museum had some displays of early rockets (likely from the Mercury program).

Actually, the space program was coming into its own in the 1960s with the Gemini and then Apollo missions sending astronauts into orbit and eventually to the moon. During the writing of this handbook, the country celebrated the 50th anniversary of Neil Armstrong, Buzz Aldrin, and Michael Collins' first flight to the moon, resulting in humans walking on its surface. I want to mention this at this point as the space program was kind of a first revelation to me how individuals, teams, and technologies work together to achieve a monumental goal.

The space program was brought front and center while I was at Rolling Fields Elementary School (Kettering) and Jefferson Elementary School (Dayton) mainly because whenever there was a space launch, teachers would bring their televisions to school and we could watch what was happening live. This brought a real aura to make science especially relevant and applicable, even more so for me. As I reflect back on these days of splendor there are some very striking foundations of my awareness. This really is the first time wherein a command post with people working together (i.e., teamwork/interaction) was shown in process. The notorious ground control function was a large command post in Houston where engineers, scientists, and communication specialists all performed under a sense of pressure (stress to complete their specific role and function as coordinated with other specialists to produce a joint outcome). And while I was aware that teamwork was important in the mission control room to make Houston happen, I was not—as a ten year-old—cognizant of the concept of team of teams. But that was exactly what was happening on space launch days. There were multiple teams within ground control responsible for specific functionalities to be achieved (sub-goals) wherein the outputs of teams were connected to input, throughput, and output of other teams to achieve major timely goals, i.e., team performance. And teams were also needed in Cape Canaveral as well to facilitate the actual rocket launch on site. But most of all the astronauts typically worked together as part of a two-person or three-person team on the space capsule itself. In the case of Apollo 11 the three-person team then bifurcated into a two-person team engaged in the control of Challenger to approach the landing on the moon, then in preparing and actually walking on the moon, but at the same time still working with Michael Collins in the main space capsule.

Not only was the prominence of tremendous teamwork on display for a nation to be in awe of, but for me this was the time when remotely located, distributed work actually meant something for a real situation. While I must say these words per se were not part of my vocabulary given my age, they were understood intrinsically through the communication and actions that emerged as astronauts engaged tasks in their respective environments. The moon is 238,855 miles from earth. Seeing astronauts that far away communicate and talk to ground control provided an applied understanding of what these words meant in reality. Not only were they communicating remotely from such a far-off distance, but remote control and monitoring of information within the spacecraft and throughout the mission demonstrated the notions of distributed remote interaction and distributed information as used across different contexts. The control dynamics associated with information, humans, and joint actions taken could be done at distance! It seemed magical for a ten year old!

The notion of distributed also connoted that the role of context was heavily transacted with the way information is perceived and used to make things happen across distance. Today we might think about this as related to ecological dimensions of time-space-control. Often this is specified by information embedded in the mutual interactions of affordances and effectivities (see Young & McNeese, 1995)—even though obviously at ten years old I did not understand these principles—they certainly apply to many complex problems in current culture. One example of this is the idea that gravity changes the time-space-control interaction that astronauts had to learn in order to move and work in space. Ironically, I remember watching simulations that they used to experience what loss of gravity would be like in space and realizing for the first time what a simulation was and how it could help astronauts become aware of their new situations (even though I knew nothing about situation awareness in 1964).

While observing that teamwork was highly resident throughout the space mission there was one additional element definitely worth mentioning. It is that the command post (mission control and launch control) was a place where technological support was critical for enabling team interaction throughout various stages of the mission. This could take different shapes. The space mission was the first place I recognized that NASA team members (albeit lined up next to each other in linear fashion) were interacting with large screen display technologies with information specific to a team's function. Concomitantly, team members also used specific control devices to manipulate information and perform physical actions requisite within their task demands. This turned out to be an early example of classic human factors—as engaged to support the overall space mission pertaining to individual, team, and team of team applications. It was relevant to me. It represented multiple layers of designs to be considered specifically in support of performance. It also showed individuals worked in the context of a team and that teams needed to collaborate with other teams to jointly produce work that accomplished tactical and strategic objectives—amazing outcomes!

In summary, this early realization provided recognition that teams were incredibly important to achieving complex system objectives and mission success. This is when teams, and in particular, distributed teams connected through technology, became real for me. Why do I report these early reflections? Because many chapters throughout this handbook take a distinct perspective and different perspectives do not just magically appear—they come from somewhere, from given influences. These innate perspectives are typically motivated by an individual's passion in life (often at an early age). That passion along with their personal mindset, interests, and experiences help to shape the research topics which constitute unique formulation about teams, cognition, and technologies. As a case in point, my early proclivities toward design and invention as related to cars, flight, and space flight eventually became the girders that produced much of my interdisciplinary research space. As irony would have it, 30 years later I would be working in the design of collaborative environments as informed by team performance experiments in the context of distributed command posts (more on that in a bit).

Imprints and Opportunity in Dayton and Beyond

While I continued my car design niche in the fifth grade (I began mixing two genres together by creating designs for rocket-powered cars and jet-powered water bikes), I also started to read more about inventors in general but specifically the Wright brothers. They actually began their expertise with bike mechanics and designs prior to developing ideas pertinent to flight. As Bernstein (1996) points out the Wrights and other inventors in Dayton were what he referred to as "grand eccentrics." Many of their beliefs and ideas were off the beaten path. Another intriguing avenue of the Wright brothers' approach was bionic design wherein they observed birds in flight and tried to emulate the processes, structure, and mechanisms of birds within their concepts for flying machines. Little did I know in the early 1960s that I would later be intimately involved in the engineering and design of human-centered aircraft at Wright-Patterson—sometimes we are surprised where life takes us and how we became scientists within team cognition. The paths and trajectories are often eccentric, nonlinear to normal progressions. While the Wright brothers are the most famous inventors and people from Dayton there are some other inventor influences that need to be mentioned, too.

Another element of technological progress intrinsic to Dayton was the National Cash Register Company (NCR) as headed up by John Patterson—he was the chief executive officer. Yes, this is the "Patterson" which Wright-Patterson Air Force Base is named after. NCR headquarters in Dayton was actually one of the first and preeminent companies involved in the design, manufacture, and sales of mechanical computers—the cash register. The mechanical computer preceded but contributed to concepts employed in the electronic computer (cash denoting numbering and counting systems and registers for storage, memory, and manipulation of numbers—all of which are important for computing). John Patterson became internationally famous not only for overseeing the NCR company and producing cash registers internationally, but as a pioneer in innovative management and merchandising techniques which contributed to the industrial revolution in America. Charles Kettering actually worked at NCR before establishing Delco and invented the first electronic cash registers while there. Eventually Kettering would become chief of research at GM where he went on to invent spark plugs, leaded gasoline, four-wheel brakes, and the automatic transmission. His home was the first in the United States to employ air conditioning (via Freon, which was invented under his leadership). Later in his life, along with GM president Alfred Sloan, he established the Sloan-Kettering Institute for Cancer Research. Truly, this was an amazing fellow and inventor's inventor.

As it would turn out, NCR and IBM were two of the very first computer-based companies to begin research and design in electronic computing, therein developing many types of computational products that are still in the marketplace today. Both companies also incorporated new avenues that helped to pioneer the field of human-computer interaction (HCI) that preceded many of the contemporary company's focus and incorporation of user experience in their product designs (see Grudin, 2005 for a history of HCI; Stuster, 2006 for stories about early human factors accomplishments; Harrison, Henneman, & Blatt, 1994 for early HCI work at NCR). An early reported extension of the NCR computer system was first used in the development of bar code scanning for supermarket checkout. These systems were developed in the late 1960s and represent some of the first intersections of humans, computers, and commercial products. The scanned checkout is still active today to help customers process their items efficiently independent of the grocery clerk. Early human factors research was conducted on this system by Lewis Hanes, NCR Research Center, Dayton, Ohio (see Stuster, 2006). The first in-store use of the barcode scanner was at the Marsh Supermarkets, Inc., Troy, Ohio. One small irony is that when my wife Judy and I were first married we lived in Troy and shopped at that very store where the first scanners were tested. Hence, the history of NCR has been very prominent in computation for a long period of time.

Therein, my life in Dayton was heavily entwined with design and aspects of how design could inspire new creations to unfold. As I spanned from grade school into high school (ironically at the John H. Patterson Cooperative High School in Dayton) my hobbies and interests turned more seriously towards design. I was accepted into the drafting and engineering specialization at Patterson. The high school was entirely unique in that it preceded what we now term career academies.[1] A student selected a career path that they hoped to pursue and coursework was designed around that trajectory. But more importantly as juniors and seniors one was able to work in business and industry in an area highly related to career choice. We would—cooperatively—work for two weeks in our jobs, then go to school for two weeks to study academic material inclusive of college preparatory and technical coursework. To do this required going to high school through July but at age sixteen I was working in the role of an engineering assistant/nameplate designer. This kind of high school experience prepared me greatly for college where I majored in engineering design and psychology.

When I graduated from the University of Dayton in 1977 I went to work as a draftsperson and worked for four years "on the board" earning my way up to senior designer. My work in this initial post-college position was actually at the Wright-Patterson Air Force Base for the Systems Engineering Avionics Facility (SEAFAC). This was critical as it exposed me very early on to systems engineering and computing, in particular avionics design. Most all of my cooperative education experiences in high school and college focused on engineering design rather than industrial design. Yet my true interests were strongly emerging in the area of design of systems that involved humans which at the time was referenced as human factors, engineering psychology, or systems psychology. In 1981, I got my big break towards that end which sparked a career that led to human factors and eventually research in team cognition.

While working as a designer at SEAFAC my employer paid for me to get a master's degree in experimental cognitive psychology (again at the University of Dayton). In 1977, I started taking late afternoon and evening coursework to accomplish this goal. I was trying to fuse together a multidisciplinary approach with some of my elective courses to forge a cognitive approach to computing and human factors (what would later transmogrify into two of my main areas, human-computer interaction and artificial intelligence). In 1981, I had all my course work completed and was beginning work on my MA thesis which was focused on the perceptual recognition of complex images using hemispheric specialization (traditional cognitive psychology experiments were conducted). One day I was out for a walk and had lunch in a scenic area at Wright-Patterson Air Force Base. I had however forgotten my drink. After I ate I was thirsty so I decided to go to the nearest building for a drink of water or a vending machine Coke. I walked into a building nearby and as I looked up in the corridor there was a sign that said "Human Factors Branch." To make a long story short I ended up interviewing for a civil servant position—engineering psychologist—in that branch and in June of 1981 I became an employee of the USAF! My early childhood dreams of designing aircraft were beginning to come true in an unusual way to say the least. I worked for my branch chief, Dr. Richard Schiffler, in the home office for a few years and was assigned also to the KC-10 systems program office to be the human factors engineer for that entire program.

Another big break came along in 1983 when I was assigned to the "skunkworks" (USAF Crew Station Design Facility) to be part of the design team of the electronic, computerized cockpit for the F-16. This was a dream job as there were many simulators there and in particular an F-16 hybrid that was hooked up to PDP-11 computers to generate imagery-symbology and new ideas for informatics in the cockpit. This enabled running many experiments utilizing novice and expert pilots involving human-computer interaction designs for the F-16 cockpit. After the pilots would participate in their experiment we would conduct in-depth interviews with them, and then go back to the drawing board to incorporate changes in design for the next experiment to continue to test performance to the point where it showed improvements. This process enabled a form of continuous process improvement that eliminated errors and created the best possible human-computer interface for the pilot. This is about as good as it gets for a human factors career in applied research and design. However, I was missing a key critical piece—the basic science component of human factors, cognition, and design. Therein, in 1984 I did a lateral transfer to the USAF Aerospace Medical Research Lab (AFAMRL) as an engineering research psychologist. I am sorry to belabor one with all the messy details to get to this point but this was the developmental history and awareness that brought me right up to the point of engagement with distributed cognition/team cognition/information communication technologies.

One of the things that I learned working in real-world human factors is that the power of the context often drives decisions as to how to support the human (or a team) within complex systems design. That has certainly been the case as my research area translated from applied human factors engineering to research in team performance, and is often specifically the basis of understanding what distributed team cognition means. The context of teamwork has driven how my interests and perspectives evolved over time. Context can infer situated settings and fields of practice and points to the socio-technological sense surround that an individual or team encounters. I was first intrigued with context and what it meant, with interest in space operations as a child (e.g., what it was like flying in a spacecraft getting ready to land on the moon). But the depth of meaning progressed to teams in support of various fighter pilot operations. While at work in ASD I soon discovered that most jobs were not really only individualistic but touched on team operations at different points in time.

When I became a scientist at AFRL in 1984 the first focus of my research was looking at how cognition facilitated individual and team performance. With respect to teamwork, the context focused on military command, control, communication, and intelligence (C^3I) communities. Part of this research thrust included how to design support systems for teams that usually involved problem solving, decision making, situation awareness, and actionable intelligence. This was literally the point in my timeline when research in team cognition began (1984). Research began attuning the use of new information and communication technologies for select tasks one would encounter in these environments (e.g., large group display to display information for use in military command posts, McNeese & Brown, 1986). Therein, one could frame this as principles of human factors engineering applied to a specific context that included the presence of dynamic teamwork. It signaled the "coming of age" of my intellectual interests and passions with my specific academic training. While this was a kind of "basic level" approach taken in the mid-1980s the research specificity morphed and sharpened into looking at how the cognitive components of teamwork (team cognition) might be amplified and supported by intelligent systems (e.g., decision aids). The interaction of team cognition (human and machine agents) within specified C^3I environments revealed challenging dynamics and wicked constraints, distributed and frequently only partial information availability, and the distributed presence of human actors outside of a given locale. It necessarily included mutual levels of influence from tangible variables such as individual differences, cultural predilection, and temporal integration/rhythm, to name a few. This interaction thus chiseled a foundation for what we term distributed team cognition.

So why was this time in the USAF important in contributing to my awareness (beyond what I have communicated above) in reference to distributed team cognition? First, it melded together my passion for technological design and psychological science. Second, it made an amazing connection between basic research and applied real-world settings (what the government/military referenced as 6.2 applied research levels of R&D) while still affording some focus on 6.1 basic research, which I considered the perfect mix. Third, it amplified my ability to actually conduct interdisciplinary research according to my joint interests in cognition, computer information sciences, human factors engineering, and collaborative systems and technologies, and to publish in these areas. Fourth, it enabled me to communicate and work with world leaders in the above-mentioned areas and to understand their collective expertise (renowned scientists and engineers such as Mica R. Endsley, Herbert Simon, Robert Williges, Kenneth Boff, William Rouse, Gary Klein, Jens Rasmussen, Eduardo Salas, John Flach, Kenneth Hammond, David Woods, Robert Hoffman, Penny Sanderson, and many others). And the connection at AFRL also provided some of my closest research colleagues (some of whom I still work with). My experience at AFRL in C^3I more specifically allowed me to look at the interconnections between individual and team performance (as relative to command posts at first but then more generally later on) and how this performance could be supported through collaborative systems—information and communication technologies. Finally, and perhaps most importantly, it provided two educational opportunities that significantly amplified my career in distributed team cognition: (1) progression towards a PhD (through USAF support) at Vanderbilt University in cognitive science (1989–1992) and (2) provision of a sabbatical taken as a visiting professor at The Ohio State University in the Department of Integrated Systems Engineering (1997–1999).

While at Vanderbilt University I directed half my focus and coursework towards computer science and AI, allowing more technological insights to flourish. As part of the program I also emphasized the areas resonant within cognition (memory, learning, problem solving, decision making, collaborative work) which would continue to guide my research niche many years into the future and in fact formulate the foundational edge for this handbook. This allowed me to do research in cooperative learning, artificial intelligence, decision aiding, human-computer interaction, and computer-supported cooperative work. However, while at Vanderbilt I received more exposure to other philosophies in cognitive science which led to even more awareness of the ecological approach to cognitive systems. This ended up creating—for me anyway—a belief and philosophy in what is best described as distributed cognition, situated cognition, or embedded cognition. One of the chapters in this handbook is by my office mate at Vanderbilt, Dr. Michael Young (and his colleagues). We both grew up under the intellectual influence of Dr. John Bransford which influenced our mutual thinking around distributed/situation cognition and the practice of learning. In 1995, we jointly published a book chapter that specifically focused on what a situated cognition approach would mean for the area of problem solving (Young & McNeese, 1995). You will see visions of this in our chapters and elsewhere but the handbook is designed to present multiple philosophies so there are many approaches present. But this is the approach I have taken and continue to advocate.

As is often the case in government research laboratories one can be reassigned to promising new areas to generate new insights and research responsibilities. Once I returned to Wright-Patterson AFB from Vanderbilt I was placed in the position of Director of the Collaborative Design Technology Laboratory. I was in charge of studying how collaborative technologies could help improve performance in distributed design teams (as a field of practice). This is a great example where distributed team cognition was present. In my vast amount of years as a researcher this was really the only cognate area where I was both an actual practitioner and researcher which obviously made it highly unique (e.g., McNeese, Zaff, Brown, Citera, &

Wellens, 1992; McNeese, Zaff, Brown, Citera, & Selvaraj, 1993). Stated another way it represented the apex of my passion for design and my increasing proclivity to study team cognition wherein collaborative technologies could make a difference in team performance. We really looked at a variety of variables within distributed design teams using different kinds of methodologies and approaches (see Citera et al., 1995; McNeese, Zaff, Citera, Brown, & Whitaker, 1995; Whitaker, Selvaraj, Brown, & McNeese, 1995).

In the mid-1990s, in yet another twist of positions at AFRL, after this first lab went away I was made director of a bigger laboratory, the Collaborative Systems Laboratory, which crossed team cognition, information and communication technologies, and situated context. This newer laboratory returned the contextual focus back to C^3I and focused on looking at more generalizations with respect to distributed team cognition. These laboratories and the research conducted therein represent the interdisciplinary, transformative nature of my research direction and passions.

As the millennium came to a close major new changes were in store for my career. Based on the research established at AFRL in team cognition and cognitive science I took a job as associate professor in the new School of Information Sciences and Technology at The Pennsylvania State University (University Park, Pennsylvania) where I stayed for 17 years until my retirement in 2017. (I am currently Professor Emeritus at Penn State University.) One of the junctures representing the first half of my research career was the publication of our *New Trends in Cooperative Activities* book in 2001 (McNeese et al., 2001). As mentioned at the beginning of this chapter that was two decades ago and a lot happened at Penn State while I was there. I continued publishing and getting various grants associated with distributed team cognition. I was able to translate doing research in the C^3I context (from the military arena) into the field of practice of emergency crisis management (civilian sector) in additional to a few other contexts.

As mentioned one can have many directions and focuses when studying teams in multiple contexts under varying conditions. The research our groups have conducted over the years has changed quite a bit and been readily modified dependent on the specificity of details within the boundary constraints of the context studied. To be more concrete early on in the USAF (mid-1980s) the context focused on military and primarily USAF command posts but this morphed more into emergency crisis management during the 2000s. There are similarities but also many differences that abound. After 9/11 the surge of interest in research related to crisis operations and crisis informatics became much greater and many of us were motivated to understand how teamwork contributed to contemporary crisis management problems in order to improve real-world operations (Brewer & McNeese, 2004; MacEachren, Fuhrmann, McNeese, Cai, & Sharma, 2005; McNeese et al., 2005).

An Emergent Research Pathway

As an example of our research profile in emergency crisis management, one of the specific research paradigm-contexts that we have used extensively is called the NeoCITIES simulation.[2] The context underlying NeoCITIES conveys a notional representation of a 911 call center dispatch engaged in emergency crisis management operations as a team (Hamilton et al., 2010; Jones, McNeese, Connors, Jefferson, & Hall, 2004; McNeese et al., 2005). While the representation is not real per se it is built on knowledge elicited from real experts and insights from observations in the field (Terrell, McNeese, & Jefferson Jr, 2004). The contextual foundation of the simulation is predicated on the idea that a functional crisis management team consists of fire, police, and hazmat members that work both together and separately. They need to jointly address team decision making while being simultaneously engaged in individual decision making as relevant for their respective tasks. The simulation incorporates events and situations that emerge across time which have differing demands. The demands that arise require resources unique to each member, and that have to be allocated at strategic points in time to reduce the severity of an event (e.g., the fire member allocates the resource fire truck and the hazmat member allocates a cleanup crew to jointly respond to the situation). Team resources are limited in amounts and time they stay at an event, therein the allocation decision needs to be made tactically but in the midst of the emergent joint situation. The team cognition required employs distributed information that helps to define the problem state and team situation awareness, and leads to correct decisions about action. In this problem, the team must understand the overall problem as it emerges, be aware of joint demands (team situation awareness), while still devoting part of their attention to individual tasks that occur as well. While the situations are fairly well-defined the team must plan, share, and communicate to be successful (emulative of the real-world team activities).

Inherent in the NeoCITIES simulation is the idea that team members could be distributed across space (remotely located) but yet have the demand to work synchronously. NeoCITIES is an example of what a research paradigm-context can be for distributed team cognition, as defined in one particular way, but is just one instantiation of what could be established. The chapter by Marhefka et al. (in this book) provides more in-depth research using NeoCITIES and clarifies experiments with more specificity. As one reads through this handbook other unique research paradigm-contexts will be provided to help a reader expand their knowledge of research within distributed team cognition.

CONCLUDING REMARKS

Many of us have come in contact with the themes provided in this first chapter (teamwork, cognition, technology, context) albeit in unique and very different ways. I have just reviewed my own personal story in terms of developmental and cultural derivation as it relates to becoming a scientist exploring distributed team cognition. Each author within this handbook could do something similar. As you read through the contributions presented keep in mind that each set of authors address their respective subject as a function of their own personal history and a set of experiences. While this produces multiple perspectives that help to explain the values of distributed team cognition, the integration of perspectives traces invariance in a way where at given levels of abstraction, common ground is certainly possible to a degree. As the next decade of research emerges, new perspectives about truth will dawn as the boundaries between human and artificial intelligences are lessened.

NOTES

1. In 1968 the John H. Patterson Cooperative High School was designated as a nationwide top ten high school primarily based on the innovation in cooperative education at the high school level.
2. NeoCITIES was conceptually derived from the original CITIES simulation, see Wellens (1993).

REFERENCES

Bernstein, M. (1996). *Grand eccentrics: Turning the century: Dayton and the inventing of America*. Wilmington, OH: Orange Frazer Press, Inc.

Bransford, J. D., Brown, A. L., & Cocking, R. R. (2000). *How people learn: Brain, mind, experience, and school* (Expanded ed.). Washington, DC: National Academy Press.

Brewer, I., & McNeese, M. D. (2004). Supporting work in hurricane management centers: An application of cognitive systems engineering techniques. In *Proceedings of the 48th annual meeting of the Human Factors and Ergonomics Society* (pp. 2426–2430). Santa Monica CA: Human Factors and Ergonomics Society.

Brockmeier, J. (2012). Narrative scenarios: Toward a culturally thick notion of narrative. In J. Valsiner (Ed.), *Oxford library of psychology. The Oxford handbook of culture and psychology* (pp. 439–467). Oxford: Oxford University Press.

Brown, J. S., Collins, A., & Duguid, P. (1989). Situated cognition and the culture of learning. *Educational Researcher, 18*(1), 32–41.

Citera, M., McNeese, M. D., Brown, C. E., Selvaraj, J. S., Zaff, B. S., & Whitaker, R. (1995). Fitting information systems to collaborating design teams. *Journal of the American Society of Information Science, 46*(7), 551–559.

Cooke, N. J., Gorman, J. C., Myers, C. W., & Duran, J. L. (2013). Interactive team cognition. *Cognitive Science, 37*, 255–285.

Cooke, N., & McNeese, M. D. (2013). Preface to special issue on the cognitive science of cyber defense analysis. *EAI Endorsed Transactions on Security and Safety, 13*(2).

DeGloma, T. (2010). Awakenings: Autobiography, memory, and the social logic of personal discovery. *Sociological Forum, 25*(3), 519–540.

Dewey, J. (1938). *Experience and education*. New York: Macmillan Company.

Endsley, M. R. (1995). Measurement of situation awareness in dynamic systems. *Human Factors, 37*(1), 65–84.

Fan, X., McNeese, M., & Yen, J. (2010). NDM-based cognitive agents for supporting decision making teams. *Human-Computer Interaction, 25*(3), 195–234.

Grudin, J. (2005). Three faces of human-computer interaction. *Annals of the History of Computing, 27*(4), 46–62.

Hall, D. L., & Jordon, J. M. (2010). *Human-centered data fusion*. Boston, MA: Artech House Publishers.

Hamilton, K., Mancuso, V., Minotra, D., Hoult, R., Mohammed, S., Parr, A., . . . McNeese, M. (2010). Using the NeoCITIES 3.1 simulation to study and measure team cognition. In *Proceedings of the 54th annual meeting of the Human Factors and Ergonomics Society* (pp. 433–437). San Francisco, CA: Human Factors and Ergonomics Society.

Harrison, M. C., Henneman, R. L., & Blatt, L. A. (1994). Design of a human factors cost-justification tool. In R. Bias & D. Mayhew (Eds.), *Cost-justifying usability* (pp. 203–242). San Diego, CA: Academic Press.

Hollan, J., Hutchins, E., & Kirsh, D. (2000). Distributed cognition: Towards a new foundation of HCI. *Transactions on Computer Human Interaction, 7*, 174–196.

Hutchins, E. (1995). *Cognition in the wild*. Cambridge, MA: MIT Press.

Jones, R. E. T., McNeese, M. D., Connors, E. S., Jefferson, Jr., T., & Hall, D. L. (2004). A distributed cognition simulation involving homeland security and defense: The development of NeoCITIES. In *Proceedings of the 48th annual meeting of the Human Factors and Ergonomics Society* (pp. 631–634). Santa Monica CA: Human Factors and Ergonomics Society.

Lave, J., & Wenger, E. (1991). *Situated learning: Legitimate peripheral participation*. Cambridge, UK: Cambridge University Press.

MacEachren, A. M., Fuhrmann, S., McNeese, M. D., Cai, G., & Sharma, R. (2005). Project highlight: Geocollaborative crisis management. In *Proceedings of the 6th annual national conference on Digital Government Research: Emerging Trends* (pp. 114–115), Atlanta, GA: U.S. National Science Foundation and the Digital Government Research Center.

Marhefka, J., Mohammed, S., Hamilton, K., Tesler, R., Mancuso, V., & McNeese, M. D. (2020). Mismatches between perceiving and actually sharing mental models: Implications for distributed teams. In M. D. McNeese, E. Salas, & M. Endsley (Eds.), *The handbook of distributed team cognition, book 1: Theoretical and historical perspectives*. Boca Raton, FL: CRC—Taylor & Francis Publishers.

Maynes, M. J., Pierce, J. L., & Laslett, B. (2008). *Telling stories: The use of personal narratives in the social sciences and in history* (1st ed.). Ithaca: Cornell University Press.

McNeese, M. D. (2000). Socio-cognitive factors in the acquisition and transfer of knowledge. *International Journal of Cognition, Technology, and Work*, *2*, 164–177.

McNeese, M. D. (2006). The interdisciplinary perspective of humane intelligence: A revisitation, twenty years hence. In *Proceedings of the 50th annual meeting of the Human Factors and Ergonomics Society* (pp. 762–766). San Monica, CA: Human Factors and Ergonomics Society.

McNeese, M. D., Bains, P., Brewer, I., Brown, C. E., Connors, E. S., Jefferson, T., . . . Terrell, I. S. (2005). The NeoCITIES simulation: Understanding the design and methodology used in a team emergency management simulation. In *Proceedings of the Human Factors and Ergonomics Society 49th annual meeting* (pp. 591–594). Santa Monica, CA: Human Factors and Ergonomics Society.

McNeese, M. D., Bautsch, H. S., & Narayanan, S. (1999). A framework for cognitive field research. *International Journal of Cognitive Ergonomics*, *3*(4), 307–332.

McNeese, M. D., & Brown, C. E. (1986). *Large group displays and team performance: An evaluation and projection of guidelines, research, and technologies*. AAMRL-TR-86-035. Wright-Patterson Air Force Base, OH: Armstrong Aerospace Medical Research Laboratory.

McNeese, M. D., & Forster, P. K. (Eds.). (2017). *Cognitive systems engineering: An integrative living laboratory framework*. Boca Raton, FL: CRC Taylor and Francis Publishing, Inc.

McNeese, M. D., & Hall, D. L. (2017). The cognitive sciences of cyber security: A framework for advancing socio-cyber systems. In P. Liu, S. Jajodia, & C. Wang (Eds.), *Recent advances in cyber situation awareness*. New York: Springer Publishing, Inc.

McNeese, M. D., Mancuso, V., McNeese, N., Endsley, T., & Forster, P. (2013). Using the living laboratory framework as a basis for understanding next generation analyst work. In *Proceedings of the SPIE DSS conference* (p. 87580F). Wellingham, WA: The International Society of Optics and Photonics.

McNeese, M. D., Mancuso, V., McNeese, N. J., Endsley, T., & Forster, P. (2014). An integrative simulation to study team cognition in emergency crisis management. In *Proceedings of the 58th annual meeting of the Human Factors and Ergonomics Society* (pp. 285–289). San Monica, CA: Human Factors and Ergonomics Society.

McNeese, M. D., & Perusich, K. (2000). Constructing a battlespace to understand macro-ergonomic factors in team situational awareness. In *Proceedings of the Industrial Ergonomics Association/Human Factors and Ergonomics Society (IEA/HFES) 2000 Congress* (pp. 2-618–2-621). Santa Monica, CA: Human Factors and Ergonomics Society.

McNeese, M. D., Salas, E., & Endsley, M. (Eds.). (2001). *New trends in cooperative activities: System dynamics in complex environments.* Santa Monica, CA: Human Factors and Ergonomics Society Press.

McNeese, M. D., Zaff, B. S., Brown, C. E., Citera, M., & Selvaraj, J. A. (1993). Understanding the context of multidisciplinary design: A case for establishing ecological validity in the study of design problem solving. In *Proceedings of the 37th annual meeting of the Human Factors Society* (pp. 1082–1086). Santa Monica, CA: Human Factors Society.

McNeese, M. D., Zaff, B. S., Brown, C. E., Citera, M., & Wellens, A. R. (1992). The role of a group-centered approach in the development of computer-supported collaborative design technologies. In *Proceedings of the 36th annual meeting of the Human Factors Society* (pp. 867–871). Santa Monica, CA: Human Factors and Ergonomics Society.

McNeese, M. D., Zaff, B. S., Citera, M., Brown, C. E., & Whitaker, R. (1995). AKADAM: Eliciting user knowledge to support participatory ergonomics. *International Journal of Industrial Ergonomics, 15*(5), 345–363.

McNeese, N. J., & Reddy, M. C. (2017). The role of team cognition in collaborative information seeking. *Journal of the Association for Information Science and Technology, 68*(1), 129–140.

Meister, D. (1999). *The history of human factors and ergonomics.* Mahwah, NJ: Lawrence Erlbaum Associates.

Mohammed, S., Hamilton, K., Tesler, R., Mancuso, V., & McNeese, M. D. (2015). Time for temporal team mental models: Expanding beyond "what" and "how" to incorporate "when". *European Journal of Work and Organizational Psychology. Special Issue: Dynamics of Team Adaptation and Team Cognition, 24*(5), 693–709.

Perkins, D. N. (1986). *Knowledge as design.* Hillsdale, NJ: Lawrence Erlbaum Associates.

Salas, E., Cooke, N. J., & Rosen, M. A. (2008). On teams, teamwork, and team performance: Discoveries and developments. *Human Factors, 50*(3), 540–547.

Salas, E., Fiore, S. M., & Letsky, M. P. (Eds.). (2013). *Theories of team cognition: Cross-disciplinary perspectives* (Series in applied psychology). New York: Routledge/Taylor & Francis Group.

Scribner, E., & Tobach, E. (Eds.). (1997). *Mind and society: Selected writings of Sylvia Scribner.* Cambridge, UK: Cambridge University Press.

Stuster, J. (Ed.). (2006). *Human factors and ergonomics society: Stories from the first 50 years.* Santa Barbara, CA: Human Factors and Ergonomics Society.

Terrell, I. S., McNeese, M. D., & Jefferson Jr., T. (2004). Exploring cognitive work within a 911 dispatch center: Using complementary knowledge elicitation techniques. In *Proceedings of the 48th annual meeting of the Human Factors and Ergonomics Society* (pp. 605–609). Santa Monica CA: Human Factors and Ergonomics Society.

Varela, F. J., Thompson, E., & Rosch, E. (1991). *The embodied mind: Cognitive science and human experience.* Cambridge, MA: MIT Press.

Walker, A. (2017). Critical autobiographical research. *The Qualitative Report, 22*(7), 1896–1908.

Wellens, A. R. (1993). Group situation awareness and distributed decision making: From military to civilian applications. In J. Castellan (Ed.), *Individual and group decision making: Current Issues* (pp. 267–291). Hillsdale, NJ: Lawrence Erlbaum Associates.

Wellens, A. R., & McNeese, M. D. (1987). A research agenda for the social psychology of intelligent machines. In *Proceedings of the IEEE National Aerospace and Electronics conference* (pp. 944–949). Dayton, OH: IEEE Aerospace and Electronics Systems Society.

Wellens, A. R., & McNeese, M. D. (1999). The social psychology of intelligent machines: A research agenda revisited. In H. J. Bullinger & J. Ziegler (Eds.), *Human computer interaction: Ergonomics and user interfaces* (pp. 696–700). Mahwah, NJ: Lawrence Erlbaum.

Whitaker, R. D., Selvaraj, J. A., Brown, C. E., & McNeese, M. D. (1995). *Collaborative design technology: Tools and techniques for improving collaborative design.* AL/CF-TR-1995-0086. Wright-Patterson Air Force Base, OH: Armstrong Aerospace Medical Research Laboratory.

Wickens, C. D., Hollands, J. G., Banbury, S., & Parasuraman, R. (2013). *Engineering psychology and human performance* (4th ed.). New York: Routledge Publishing.

Wilson, M. (2002). Six views of embodied cognition. *Psychonomic Bulletin & Review*, *9*(4), 625–636.

Young, M. F., & McNeese, M. D. (1995). A situated cognition approach to problem solving. In P. Hancock, J. Flach, J. Caird, & K. Vincente (Eds.), *Local applications of the ecological approach to human-machine systems* (pp. 359–391). Hillsdale, NJ: Lawrence Erlbaum Associates, Inc.

Zhang, J., & Norman, D. A. (1994). Representations in distributed cognitive tasks. *Cognitive Science, 18*, 87–122.

Zins, C. (2007). Knowledge map of information science. *Journal of the American Society of Information Science and Technology*, *58*(4), 526–535.

2 Reflections on Team Simulations—Part I
Historical Precedence

Michael D. McNeese, Nathaniel J. McNeese, Lisa A. Delise, Joan R. Rentsch, and Clifford E. Brown

CONTENTS

INTRODUCTION

The research landscape to study and understand teamwork and team cognition has been complex and has evolved through various eras of influence—often determined by the zeitgeist of the times (e.g., orientations may reflect behavioral, cognitive, and ecological values of the researcher conducting the studies). In particular, the study of team cognition has become increasingly focused on distributed aspects of teams related to information, technology, people, place, and environment. In the recent years there has been a blurring between the research areas of team cognition (Cooke, Gorman, Duran, & Taylor, 2007; Salas, Fiore, & Letsky, 2012) and distributed cognition (Hollins, Hutchins, & Kirsh, 2000; Nardi & Miller, 1991), therein the *raison d'être* for this handbook. The handbook, including this chapter, illustrates how these areas meld together into the integrated research niche of distributed team cognition. We have developed two interrelated chapters (Chapter 2 and Chapter 3) for the purpose of examining a longitudinal perspective and reviewing distributed team cognition research that the authors have been intimately involved with in collaboration with the senior author over the last 35 years. Both chapters examine theoretical foundations, methodological tools, collaborative technologies, and pertinent measures that afford new levels of understanding, insight, and advancement. This chapter develops a historical precedent that lays (1) the foundation for development and (2) the conceptual underpinnings of interdisciplinary research in distributed team cognition, whereas the next chapter focuses more on a contemporary progression of research and practice.

These chapters highlight how teams and team cognition research have evolved through the use of simulations and how simulations should reflect the real-world environment they represent. In doing so we explore frequently researched topics and report what is known about these topics while reviewing lessons learned in utilizing simulation and modeling within a broader context. We also present selected preeminent simulations that have been used with much success.

As the research trajectory of team cognition is examined through various lenses associated with team simulation theory and practice, general issues may become evident as one reads through the text. There certainly are different ways to conceptualize what any team simulation is about and how it might be designed to accomplish the researcher's goals. One way to conceptualize team simulation is correspondent to Jens Rasmussen's ideas (Rasmussen, Pejtersen, & Goodstein, 1994) inherent in the abstraction hierarchy (AH). In consideration of any given team simulation it is useful to delineate the following:

- What is its purpose and how is this meaningful?
- What are some of the values that are salient?
- What abstract functions are required?
- What are the constraints and measures associated with these?
- What measurements are imminent—what measurements may be hidden?
- What priorities are suggested?
- What general functions will be needed to make it work?
- What functions will be allocated to humans, machines, and/or agents?
- What physical forms may be taken?

As scientists address these questions a simulation can be conceptualized from abstract to concrete levels of specificity, but simultaneously can be decomposed from global to local spaces. One moves through the hierarchy logically in terms of the why-what-how nature of a complex system wherein the hierarchy approaches multiple interrelated means-end relationships. That is, the most bottom levels are the means to accomplish the levels above them (their ends) which then successively become the means to accomplish the next level up and so on. When a designer or scientist builds a team simulation it may be constructed conceptually to answer these kinds of questions to comprehensively consider multiple, interrelated facets of the design. The AH provides a socio-technical perspective to understand complex systems such as simulations, and hence adds value by thoroughly considering design components when phenomena under question contain social and technical nuances. This philosophy elevates the ecological necessity underlying design, simulation, and human activities (ecological psychology, see Neisser, 1976)

Although there are many insights gained with the use of the abstraction hierarchy in conceptually designing a given team simulation, our approach utilizes a more open-ended but related proscription based on a series of questions that help define conceptualization. Our experience shows the following questions are important to consider:

1. Why is the simulation designed to begin with? What purpose does it serve?
2. What are the salient demands inherent within the context for the simulation? How could they impact human/team performance?
3. How does the simulation facilitate team cognition? What architecture and processes are utilized?
4. What are some important studies or cases that were produced while using the simulation?
5. What was learned by using the simulation? What did it produce?

Because there is a diverse population of team simulations available for specific research intentions, the focus of this chapter is from the perspective of (1) understanding cognitive processes in teamwork and how they impact outcomes, (2) extending human capabilities within teams through the use of support technologies that enhance teamwork, team performance, and communication, and (3) examining the ecological validity of a simulation within a given field of practice.

Use of Constructivism and Situated Learning

One tenet taken to assess and articulate state of the art practice is the idea of constructivism. Constructivist philosophy implies that people build up knowledge and discover new ideas from one state of experience to another (based on what they already know). Inherent in this view is that thinking and learning are coupled to the real-life contexts in which they occur—situated learning (Bransford, Sherwood, Hasselbring, Kinzer, & Williams, 1990). Hence the notion that knowledge is constructed in and through the context or a specific field of practice. Oftentimes constructivist theory is connected to ideas of distributed intelligence (Pea, 1993). Activities in problem solving require intelligence that is distributed across information, people, media, and the environment. Furthermore, constructivist philosophy often is embedded in joint social activities that provide opportunities for new learning through stages of knowledge development (Vygotsky, 1978). More broadly stated, situated cognition pertains to the perspective that people's knowledge is immersive and embedded in the activity, history, context, and culture where it was learned (Brown, Collins, & Duguid, 1989). As applied with the parameters of this chapter, an initiation point of our work within early team simulations is offered. Overall, the goal is to facilitate conceptualization by using embedded concrete examples and activities and at the same time show evolution within the research that is useful for other researchers involved with the design and utilization of team simulations.

Therein, the chapter develops by capitalizing on work from our own distributed team cognition history and the development of simulations appropriate to the needs at the time. Although the point is not to provide an ad for our own simulations it does afford fertile ground for discussing research foci, foundations, relevant issues, lessons learned, and in situ research approaches and measures. It especially offers an overall anchor point for exploring many elements of distributed team cognition in early work and providing a bridge for the follow-on chapter which in turn is more relevant to the incorporation and integration of technology, computing, and decision-making agents and aids within a simulation. Our development of simulations is indicative of the proactive use of user-centered technological innovation and design. These chapters are presented with the bias to demonstrate the symbiotic (and interdisciplinary) relationship among information, technology, people, and context. In summary, one of the main objectives is to review and show how collaborative-based simulation technologies can incorporate meaningful experiences of distributed team cognition.

HISTORY OF TEAMWORK RESEARCH AT WRIGHT-PATTERSON AFB, OHIO

It may be the case that current researchers are biased towards using literature and methods that have been published in the last five years—the recency effect. Although this is effective for having the most up-to-date findings and advancing the most precise methodologies, there is also value in knowing where research has come from (historical precedent) and why it has evolved the way it has (reasoned explanation). Our belief is that there is much value in history and that constructivist approaches utilize history to understand current context of use.

Necessity of Teams for Military Operations

Military engagement has historically involved warfighting with adversaries to obtain intentional outcomes for the purpose of securing freedom. Through the use of organizations, information, intelligence, technologies, and people, missions have been carried out with both success and failure. More contemporary military objectives include non-warfighting activities (e.g., peacekeeping operations, medical support in various endeavors, and regime restoration) and the necessary training. This being said, eventually the pathway of current practice is often traceable to objectives that involve warfare, secure operations, fighting terroristic entities, and protecting national interests and resources.

People working together effectively and efficiently as a team lies at the heart of successful operations; however this does not occur automatically. In fact, it can be challenging to take a group of people and transform them into a team that successfully accomplishes a goal with interdependent relationships. Teams formulate a core system of values that allow them to (1) strategically pull together and allocate their joint resources and (2) collectively induct their respective knowledge to perform a mission. Likewise, many operations require members to be part of different teams simultaneously and function as a "team of teams." Hence, one sees that command, control, communications, and intelligence frequently form a core set of capabilities that enable teamwork to succeed. Also, teamwork does not just happen but requires mutual cooperative learning where knowledge, skills, and abilities are acquired to perform essential tasks within a mission (Johnson & Johnson, 1994; McNeese, 1992). James Clapper, former Director of National Intelligence of the United States, mentions in his new memoir (Clapper, 2018),

> I'd been in the SIGINT world and around the rest of the community long enough to realize that each agency needed to embrace its own culture, traditions, and capabilities. After honoring that we could inspire them to cooperate to take advantage of complementary strengths . . . bringing together different perspectives and experiences enabled us to formulate a range of different options for action, . . . the old saying "the sum is greater than the parts" has profound meaning.
>
> **(p. 102)**

Inherently, the idea of teamwork contains the theme that the cognition of team members can be integrated—to be roughly on the same page—to accomplish mutual goals while yet retaining the stability of their individual performance, skills, and knowledge. When knowledge, awareness, and perspectives are disparate it is much more difficult to formulate teamwork and thereby establish correct action given the situation at hand. Clapper (2018, p. 4) refers to the creation of a new field, geospatial intelligence, as requiring the functional synthesis of mappers and image analysts, whereupon "one of our big goals was to get people with different skill sets to physically and functionally work together." In today's world, distributed work is often the norm and working together has many challenges at a distance. Many of the simulations involving team cognition therein require specification of distributed cognition—not just team cognition at collocated facilities. Distributed cognition has become possible and even an everyday occurrence. Texting on a smartphone sent to multiple parties (with videos or pictures embedded) is a very basic example of distributed team cognition. More expansive opportunities exist through the power of online collaboration spaces (that contain workflow management, discussion boards, group posts, or feeds), smartphones with various applications such as Facebook, YouTube, Google Docs, and specific collaborative software platforms designed solely to support cooperative work whether it includes video teleconferencing or group chat capability (e.g., the applications Zoom or Slack).

Within military teamwork, the role of technology is often prevalent as new technological innovations are designed and tested with real situational urgency (e.g. distributed crews working with autonomous vehicles can be used for land, space, ground, and underwater missions that typically can involve surveillance, reconnaissance, weaponization involving precision strikes, and intelligence gathering). As is often the case, technology-centered solutions may occur simply because they could be designed, and in turn fail to properly serve the human therein usually resulting in errors or wrong use. When teams and the individuals who compose teams cannot use technologies because they fail to be user-centered in their conception and design, then failure is imminent. Often times it is catastrophic failure resulting in loss of life and/or significant material goods. Therein, one of the primary reasons to begin studying teams was to eliminate errors in practice, to help facilitate cooperative learning, and to ensure technology fits the needs and capabilities of the humans involved (human factors engineering; Meister, 1999).

As one begins to study teamwork there are six major fundamental research components to consider, which have led to incisive research in both quantitative and qualitative spectrums. They are important to facilitate proper functioning and integration of teammates as they work towards mutual, interdependent activity given their select roles in the team:

1. Coordination
2. Communication
3. Collaboration
4. Cooperation
5. Awareness
6. Context/culture

These components should be taken under proper consideration to encourage and support a mutual work objective, improve decision making in cooperative work that makes a real difference, and properly design user-centered interface technologies that are successful in practice. As an example of these elements applied individually as well as collectively the reader is referred to Letsky, Warner, Fiore, and Smith (2008) and their work on macrocognition.

As history would have it, different eras of research produce related but different points of view and may even ignore other areas simply as a matter of what was salient at the time. The knowledge and culture that has informed much of the senior author's work was inspired through his work with the U.S. Air Force. In turn, this has provided a specific time and place for what was learned and practiced in team simulations and how it has evolved across the decades. A foundational conceptual question that underlies much of this research is: What is the purpose of a team simulation?

Early Beginnings in the 1960s and 1970s at USAF Aerospace Medical Research Laboratory

In order to obtain a sense of historical development of team simulations it is instructive to look at the approach, context, and outcomes that were operative and in vogue within a given phase. Therein, this section utilizes select themes to capture constructivist foundations of simulation. The senior author began work in the summer of 1977 at Wright-Patterson Air Force Base early in his career in the role of a designer. His first introduction to teams was that of participating on engineering design teams[1] that developed integrative systems avionics products within the Aeronautical Systems Division. Typically, these teams typically were composed of engineers, designers, draftspersons, technicians, and business managers. In the 1980s, he began a civil servant career applying human factors engineering to aviation systems which then changed from applied design and evaluation of real fielded products to more applied research in human factors. By way of history then, this research turned towards the path of teamwork and how teams could enhance human-system performance. As mentioned the military used teams to produce gains in operational advantage in warfare, benefits in training, and collective insight beyond what an individual could muster alone. In turn, much of the early work enabled a lifetime of research that began with practical problems found within the context of the U.S. Air Force, in particular at the Fitts Human Engineering Division, Aerospace Medical Research Laboratory, Wright-Patterson AFB, where he worked as an engineering research psychologist beginning in the mid-1980s. However, teamwork and the value of team performance were actually studied much earlier than this at the USAF Aerospace Medical Research Laboratory during the 1960s (and probably prior to this although the history of this paper will only span into the 1960s).

Through the early work of Paul M. Fitts and his colleagues (e.g., Dr. Walter Grether and Dr. Mel Warrick),[2] human factors research took root specifically at Wright-Patterson Air Force Base in the Aero Medical Laboratory. This occurred within the crossroads of military psychology practice (from the 1940s and before, see Alluisi, 1994), aviation psychology, and the subsequent emergence of the field of engineering psychology as required for military operations to have proper training and selection, avoidance of human error, and establish peak human performance (Grether, 1995). In these early days, human factors primarily focused on individual performance but not completely. The earliest team-based work in the 1960s could be thought of as an extension of experimental psychology (see Chapanis, Garner, & Morgan, 1949) applied to group settings such as air crews (see Williges, Johnston, & Briggs, 1966 for exemplary work on this) prior to the introduction of contemporary cognition studies per se. Hence, as an example some early ideas of teaming focused on how social theories might impact structural relationships in a team (Morrissette, 1958), how team size and communication (Kincade & Kidd, 1958) affect decision making, and how confinement and sustained operations impact group function (Alluisi, Chiles, Hall, & Hawkes, 1963). Even at this early juncture, there was a recognition and opportunity to address human problems with (1) theories associated with social behavior and (2) the discipline of human factors engineering (see McCormick, 1957; Meister, 1999) and hence focusing on individual-to-team performance (e.g., Lorge, Fox, Davitz, & Brenner, 1958). The signal of human factors involvement suggests that work at this time touched upon technological support albeit more primitive formulations related to equipment design.

These early simulations were more primitive and patterned after the typical toy tasks used in experimental psychology but generalized to team-level functions. Yet the research emphasized really important considerations that still remain salient for contemporary researchers engaged with the design and use of team simulations such as ecological validity, fidelity of the task simulation, level and degree of training, reliability and validity of team measures, the structure of a team, and apropos statistical analyses. Although cognition was not a foreign concept (see "cognitive dissonance," Festinger, 1962) the way it was framed prior to the cognitive revolution in psychology (see Gardner, 1985) was coupled to things like sensory-perception integration, decision making, and the level of demands inherent within a job (i.e., workload, see Kidd, 1961). Likewise, these studies were conducted prior to a strong and relevant focus on team cognition, which would come later. These early attempts to garner the power and utility of team performance for military teams laid a solid foundation for work that would emerge in the 1980s.

RESEARCH IN GROUP DECISION MAKING/TEAM PROBLEM SOLVING: SIMULATIONS IN THE 1980S/1990S

Continuing at Wright-Patterson AFB, the senior author transferred into the Aerospace Medical Research Laboratory/Human Engineering Division in 1984. The focus evolved from the early days of human factors and the effectiveness and efficiency of teamwork was garnering support as a major research topic of interest for the USAF and other branches of service. Three primary reasons for the increasing importance of teamwork were (1) command posts in real warfare operations, (2) command, control, communications, and intelligence (C^3I) operations, and (3) air crew performance and associated team training. The first author transferred into the Crew Systems Branch which had a purpose of looking at crew systems performance but readily identifying support technologies that would enhance warfighter capabilities, focusing more on areas 1 and 2 above rather than area 3.

It is worth noting that the context at this point in time drove the development of team performance out of necessity as teams were required to assess, process, interpret, and act on different kinds of information and intelligence. A popular early model used to capture this was Joe Wohl's SHOR framework (Stimulus—Hypothesis—Options—Response) model (Wohl, 1981). Although context drove the need for understanding teamwork and team performance, most approaches were still heavily coupled to quantitative, experimental approaches to understanding behavior. One of the most difficult problems to overcome was a bias towards technology-centric designs to support teamwork (the "build it and they will come" mentality) which elevated the idea that just because something could be produced it would inherently be good. The fallacy of this logic is that technology was produced but it created human errors and failure as it was designed devoid of understanding human constraints and capabilities. Therein, human factors[3] had a real calling to improve designs for individual and team performance. And herein was where the role and importance of team simulation came to pass. Team simulations could actively identify individual and team errors for various degrees of difficult tasks and pinpoint where failure would be most inclined to happen, while identifying potential causalities. Hence the value of team simulations began to be accepted.

During the 1970s the cognitive revolution took hold with much zeal and there was much more focus on how humans used their abilities such as attention, perception, memory, language, judgment and decision making, problem solving, reasoning, and learning (Gardner, 1985; Neisser, 1967, 1976). While cognition was coming into its own the approach primarily utilized the same experimental psychology paradigm and the essential elements mentioned in the last section were still relevant for designing and implementing a sound experiment. Teamwork and team performance studies started looking at cognitive activities that coupled the individual with teamwork. Therein, team simulation was often at the heart of these studies to enable controlled experimentation with high reliability and validity, under precise conditions, where multiple measurements could be acquired of participants. One major difference owing to the advancement of both software and hardware technologies (i.e., computing) was that the tasks an experimenter used were not just toy tasks but more along the lines of what has been termed a synthetic task environment (Martin, Lyon, &

Schreiber, 1998). These new computer-enabled environments provided significant power increases in display, control, precision, measurement, and flexibility while abstracting out important elements from the real-world context (Cooke, Rivera, Shope, & Caukwell, 1999). Moreover, they also enabled incorporation, testing and evaluation of collaborative technologies one might be designing and building (e.g., a large group display). Next, we examine some of these simulations that were important for our research and development.

C^3 Operator Performance Engineering (COPE) Program

The COPE program at the Fitts Human Engineering Division (see McNeese and Brown (1986) for representative work applicable to military command, control, and communications (C^3)), provided the first real opportunity for research engagement with teams especially given how team performance could be improved through types of collaborative technology and with the application of human factors engineering. This particular program allowed connection with a U.S. working group called the DAWG (Decision Aiding Working Group). This is important as it underlines the focus from earlier work that related to decision making as a major theme that continued and led to the development of team cognition as an important concept in current research. The time period representative of the COPE work is approximately 1983–1989. This work approached understanding of team interaction from a human information processing perspective (Lindsay & Norman, 1977) while still preserving the experimental psychology perspective. This is still an acceptable viewpoint in teamwork literature (Hinsz, Tindale, & Vollrath, 1997; Mesmer-Magnus & DeChurch, 2009), however it would prove to have some limitations as will be pointed out in the next chapter.

As the COPE research group considered the contexts within C^3 where cognition was pertinent for successful teamwork, it was necessary to conceptualize (1) the research purpose of a given team simulation, (2) the kinds of cognitive teamwork required and hence the specific demands inherent within the task simulation to be authentic to specific research needs, and (3) the kinds of technologies that might support cognitive teamwork. One of the pertinent frameworks that has allowed development of different simulations relevant to points 1, 2, and 3 is McGrath's Group Circumplex (1984). The circumplex breaks down task demands into specific categories as related to how a team performs. In particular, the vertical dimension of the framework looks at whether the teamwork is representative of collaboration, coordination, or conflict resolution, whereas the horizontal dimension portrays activities as either cognitive or behavioral. These crossings then result in eight different task demands. Simulations therein should represent a given kind of task demand that will help accomplish research objectives. This relates to what we mentioned in the introduction about how a researcher needs to spend copious amounts of time focused on the conceptualization of the team simulation that is appropriate to their needs. As team simulations are reviewed—emergent from the COPE program and beyond—answers to specific questions of conceptualization can be presented as apropos. As a guide, most all of our research tasks are related to decision making or problem solving therein falling within the quadrants of "generate" and "choose." Breaking that down further the tasks designed within our simulations fell into the regions of "intellective" or "creativity" or "planning" tasks.

During the late 1970s into the 1980s there was a distinctive shift in experimental psychology towards utilizing cognitive constructs as a means of understanding human behavior (the cognitive revolution). This perspective also applied to understanding team behavior and whether teams were effective or not. This constituted the human information processing perspective but it also included classical decision-making approaches such as the judgment and decision-making perspective (Baron, 2004; Tversky & Kahneman, 1974) where models portrayed optimal decisions based on laws of probability. Some of the work related to heuristics and biases eventually led in part to a new approach termed naturalistic decision making which demonstrated actual decision makers in real environments do not often utilize optimal strategies but construct naturalistic ones (Klein, Orasanu, Calderwood, & Zsambok, 1993). Many of the topics of interest to the DAWG included looking at teamwork in terms of biases, belief functions, probabilistic reasoning, heuristics, and judgment. Consequently, the development of systems and technologies to support teamwork were predicated on these kinds of perspectives coming into play, especially when complexity existed. An early example of this genre of research was conducted early in the COPE program that is indicative of looking at how decision making could be enhanced through the use of decision aids (Eimer, 1987). Dr. Erhard Eimer was an experimental psychologist and professor at Wittenberg University, who was on sabbatical at Wright-Patterson Air Force Base. Dr. Eimer provided a niche for this kind of quantitative work in team decision making study from the tradition of Kahneman, Tversky, and others, therein laying some of the groundwork that helped establish this area in the laboratory.

The earliest seminal paper (McNeese & Brown, 1986) that the senior author produced while in the COPE program focused on the use of large group display technologies in terms of how they affected team performance. This DTIC technical report looked at guidelines, research, and human factors considerations that would improve teamwork. The simulations developed under COPE were developed with the C^3 context in mind. This period of growth shows a distinct translation from the earlier focus on equipment design for teams into collaborative support systems (aids, associates, interfaces) that assist team members along the cognitive dimension. Therein, the COPE simulations developed tasks with inherent demands that centered around a specific type of decision making.

A significant component of the COPE program worked directly with national-level command posts (e.g., North American Aerospace Defense Command—NORAD) for performance improvement vis-a-vis adoption of a new technology area—human-computer interaction (Myers, 1998). This was the beginning of qualitative work representative of naturalistic decision making that helped inform understanding of missions-scenarios-problems. Research findings showed evidence that team workers felt pressure with very demanding tasks that were ensconced within coordination, communication, collaboration, cooperation, awareness, and cultural imprints (Cannon-Bowers & Salas, 1998).

Next we examine specific team simulations developed under the COPE program, and what they provided in terms of building blocks for team cognition research today.

As previously mentioned this is only one strand of team simulation work derived from the 1980s and 1990s that afforded investigation of team cognition. In order to compare/contrast any team simulation we propose using these attributes to assess the distinctiveness, power, and viability of the simulation for research purposes.

The Team Resource Allocation Problem (TRAP) was the first team simulation (introduced by Brown & Leupp, 1985) produced for the COPE program. The design and utilization of TRAP was overseen by Dr. Clifford Brown, one of the authors of this chapter. Dr. Brown, following in the footsteps of Dr. Eimer from Wittenberg University, was a National Research Council Research Associate (twice) at Wright-Patterson Air Force Base where he began work with the COPE program. During this time TRAP was designed and utilized for experimental research purposes and integrated into experimental requirements. TRAP represented a highly quantitative approach to distributed team cognition. It was a mathematically based team decision-making task that was designed to be a generic testbed for studying issues of importance in actual command, control, and communication environments. It provided multiple objective measures of team performance that could be compared across experimental conditions and compared to models of effective team problem solving. Importantly, it could be sped up or slowed down to represent the effect of time stress without altering the problem or its corresponding measures of performance. Therefore, for example, the effects on team performance of variables such as graphic versus alphanumeric display of information under low or high time stress could be systematically investigated.

The research goals using the TRAP simulation were to explore the intersections of team information display, social/organizational psychology, and cognitive processes, and to explore useful support mechanisms with specific kinds of collaborative technologies. This interdisciplinary approach became the hallmark for most of the simulations our research group has undertaken. TRAP provided a simulation indicative of interdependent, dynamic decision making. The original TRAP was developed by upgrading a complex, individual decision-making problem/task (Pattipati, Kleinman, & Ephrath, 1983) to generate a group-level problem where cognition was shared across a small team. It comprised elements of intellective choice tasks and planning tasks from McGrath's framework.

The context underlying the creation of TRAP focused on analytical reasoning, information sharing, and team collaboration within C^3 domains. These are real-world factors that come into play in actual command post interactions. Team performance is accomplished dependent on how well these activities are done. Note that team performance is a joint product of individual tasks accomplished in addition to the team-level tasks that need to be processed. Part of the complexity involves figuring out how to work individually while also interacting with team-level demands, and what takes precedence or priority at any given point in time. Some of the major principles that contributed to the dynamics of TRAP were: (a) how difficult the demands were across a small team at any given juncture (individual and team workload, urgency); (b) awareness of other team member's activities and whether they could contribute to a solution at a given point in time (information sharing and team situation awareness); (c) deriving team solution tactics with specific rules embedded in the game—with given input parameters under changing conditions (cognitive analytics and interdependency); and (d) comprehending how team members communicated and made decisions while being supported with collaborative technologies (e.g., this continued work comparing small/large group displays, see Wilson, McNeese, & Brown, 1987). TRAP represented an abstract formulation of team resource allocation within a typical C^3 context as performed in a synthetic task environment (simulation). One of the benefits derived was that individual and team performance could both be evaluated at any point in time as a team member could be working individually or as part of a team. Individual and team performance could be calculated (team performance scores were determined based on the processing of numerous tasks by the team and the point values derived). Although the context roughly approximated team cognition in a C^3 domain, the task resonated more towards abstract planning and reasoning rather than a concrete, situated team-level task that replicated an actual C^3 team task (i.e., realism). This enabled a broad framework for testing different theories and hypotheses relevant to the research base. This reified the experimental psychological/experimental design values of controlled repeatable performance trials, training to criterion levels, precise measurements, and a focus on internal validity.

Inherently, this task required timely coordination and communication to enable optimal team performance. Interdependency—one of the important principles underlying team cognition and decision making—is a prominent part of the task structure. The simulation structure enabled abstract "processing of tasks" according to: (1) whether there is sufficient time to complete the task, (2) whether the required resources are available to "process" the task, and (3) whether the point value as derived from task characteristics (e.g., shape and color) optimizes team performance. Specific tasks are worth more than others according to a defined rule set (see Brown & Leupp, 1985 for specific rule structure). A team must work through the cognitive analytics necessary to produce the highest number of accumulated points based on a combination of processing opportunities. Because team members only have a set number of resources, they need to make timely and thoughtful decisions with teammates to accrue the most points. Also, note that a team member is required to stick with a given task for the duration of the task cycle required (e.g., 15 or 30 seconds) before starting on a new task. This requires communication, coordination, task-team member awareness, and recognition of best possible outcomes by all team members (team situation awareness).

TRAP employed a three-person team working at a console wherein they have a display that shows the opportunity window with the number of tasks available within a given period of time. As time progresses in a trial different tasks (opportunities) become available for team members to process (some are for individuals, some are for dyads, and some are for all three team members). The initial setup used computers to generate the tasks on the opportunity window for a given trial and to collect data from team members for that trial. The display provided all the information required for the team to work through the tasks. Each team member utilizes a control unit (module) to move their cursor over a task; once all required team members' cursors are over the task, any one of them can press their start button to begin processing it. While processing one task, team members can move their cursors to the next task they plan to process, but until the previous task is completed (or aborted) they are not able to start processing the next task. Therein, the simulation consists of a computer system that facilitates the appropriate control and displays for the team and records data as needed. Variations were made to the display units as the information could be displayed on a team-level large group display and/or optionally on individual display units for each team member.

The value produced by processing various one-person, two-person, and three-person tasks dynamically changed across time, requiring adaptive and coordinated responses for the best overall performance solution. TRAP highlighted temporal dynamics, interdependence, and adaptive changes among team members.

One of the positive lessons learned was that designing a highly controlled team-level decision-making simulation with easily changed parameters yielded many advantages in terms of creating experimental designs and testing different hypotheses. Also, the ability to seamlessly integrate new technologies within the team simulation provided researchers with an initial ability to assess the value and impact of innovations and designs upon team performance. Another advantage was that the problem involved both individual and team performance, and the demand to switch from individual to the team focus during the course of a session (which often replicates real-world demands in decision making and problem solving). The TRAP task generated a high level of interdependency among team members which is important for simulations that hope to replicate teamwork with common goals and mutuality. Although the TRAP simulation provided a testbed for investigating variables of importance in the C^3 domain, it was abstract and generic (an experimental strength), but without any real-world context. With the growing influence of naturalistic decision making and ecological validity, TRAP was seen by some as a just a simple game. The decision making was well-defined once mastery of the decision rule set was fluent and if participants communicated at the proper times and in a clear way. This well-defined component of decision making may be an advantage, but many situated problems that occur in C^3 are ill-defined and without objectively identifiable optimal solutions.

TRAP captured some of the dynamics and coordination issues that are important for team cognition but it also had some challenges. TRAP emulated the cognitive psychology representative of the 1980s, but adapted it for understanding team performance. Given the requirements it was designed for, it provided a reliable team simulation that enabled performance with and without specific technologies, thereby enabling meaningful evaluations of their effects on teamwork

Many useful results were produced by TRAP. The first experiments investigated (1) relative differences in team performance when using small versus large group displays, (2) whether differing formats of information display (graphical versus alphanumeric representation) provided to a team helped facilitate team performance, and (3) how information presentation rates (moderate or fast) impacted team performance. The result provided useful understanding in coupling technologies with information processing for individual and team performance. Performance was also assessed with respect to subjective workload measurements showing how team information processing could impact one's perception of workload in a cognitively engaging team task. Workload during the 1980s was a measure of prime interest in applied cognition and human factors, hence one of the reasons why it was employed as a dependent measure in some of the TRAP experiments. Additional research using TRAP (see McBride & Brown, 1989) demonstrated that providing decision heuristics enhanced team performance.

C³ Interactive Task for Identifying Emerging Situations (CITIES)

The second major team simulation developed under the COPE program was C³ Interactive Task for Identifying Emerging Situations (CITIES). Dr. A. Rodney Wellens was its creator and implementer (during his time as a senior scholar at the Air Force Office of Scientific Research at Wright-Patterson Air Force Base). Dr. Brown, Dr. Wellens, and the senior author of this chapter worked together in the COPE program in the 1980s to understand the intersection of team decision making and potential use of technologies to improve performance.

As one looks at the limitations of TRAP, they provided inputs for the design of the next team simulation. The goal was to look at interdependent teamwork that emerged in a situated C³ context. In particular, the focus remained team decision making, but the purpose of the CITIES simulation was to make the context more prominent in terms of task content, materials, and interfaces that represented a real-world domain. This was directly in contrast to the abstract context-independent nature of TRAP. Hence CITIES still enabled a controlled experimental task with a requisite team composite score that required individual and team performance, dependent upon what events transpired.

Additionally, the simulation was designed to test a specific hypothesis of team-technology interaction, termed *psychological distancing* (see Wellens, 1990), wherein different kinds of technological media facilitated degrees of psychological closeness (e.g., face-to-face communication provides a high degree of closeness whereas computer messaging does not). Originally, this construct was portrayed as a linear dimension and varied from face-to-face to two-way TV (to emulate video teleconferencing) to telephone to two-way messaging (to emulate computer communication). Experiments using CITIES could assess how technology impacted closeness and in turn how this might influence team performance. Furthermore, the initial experiments pioneered the use of expert systems as a means of aiding team decision makers, and additionally pioneered the use of "talking heads" to represent how an expert system could interact with a human, in contrast to the typical text-based verbal communication (Wellens, 1993). These technological innovations made the CITIES research very much ahead of its time.

One of the strategic moves with CITIES was to generate a realistic C³ context that could be used without concerns for classified materials. Because of this constraint it was decided that the simulation could utilize emergency crisis management a context similar to C³ wherein teamwork and sharing of information for different situations/ events could be easily utilized. The overall simulation, like TRAP, was a resource allocation task, but it was more realistic. Information regarding events that took place in a city were sent to two different dispatchers for processing: one dispatcher controlled police and towing resources while the other dispatcher controlled fire and rescue resources. One element of the context that was specifically emphasized and remains of interest was team situation awareness (SA). At the time SA was introduced as an individual cognition concept (Endsley, 1995) and had not been applied much to teamwork. Therein, the CITIES simulation was unique in that it was looking at SA at the team level with the purpose to see how team SA and performance might vary with the use of different technologies, defined by their relative degree of psychological closeness (Wellens, 1993).

CITIES was similar to the abstract TRAP task in that it was predicated on team resource allocation, required individual performance as well as group-level performance, required communication and interdependent decision making, included a built-in team performance score, and employed teams to process shared information resulting in decisions that had consequences on upcoming choices and outcomes. However, the simulation was entirely embedded in real-world situations that contained events that could reveal underlying scenarios and attributes. As events unfolded greater SA across the team was possible which theoretically provided a greater opportunity for understanding what was going on and in turn what was required for resource allocation. The task required the police/tow dispatcher and the fire/rescue dispatcher to allocate specific but limited resources to be applied to events that emerged as portrayed on a computer interface. Depending on the severity and type of event, different resources would need to be allocated. If appropriate resources were not allocated to an event in a timely manner, an event could grow worse over time (e.g., a fire could grow out of control, a traffic accident could create gridlock).

Team cognition therefore involved communication among the team members to coordinate resources allocated for specific events, to keep track of them, and to keep monitoring the situation for new events and what would be needed (anticipation). Team members also had to keep an eye on whether their own resources were nearing depletion. Obtaining team SA helped to understand the big picture and know in advance what the projected resource demand and allocation would entail. It was hypothesized that different kinds of technologies (basically looking at electronically mediated communication) would afford relative levels of psychological closeness and team SA, resulting in improvement or decline in team performance scores. In summary, this simulation focused on the team cognition constructs of team resource allocation-based decision making and team SA, but was designed to gain an understanding of whether technology would either support or detract from team performance, in contrast to face-to-face performance.

CITIES utilized a fairly basic computer hardware setup of two Apple II computers comprising two experimental rooms connected to each other through a control room (Wellens, 1990). The computers presented information at each team member's workstation as the information flow dynamically changed and propagated various event streams. Standard programming was used but it was somewhat ad hoc (i.e., the programming was specific to this simulation only) for many tasks. The architecture needed to simulate and introduce an electronic aid as a team member, wherein an expert rule-based system was produced to enable team members to work with a decision aid. Uniquely, the system employed a message transformer whereupon a message from the aid could be communicated to the other team members via text-based dialogue or through a talking head which could be either a male or female representation. Furthermore, the computer systems underlying CITIES afforded great integration and depth of measurement. In addition to the composite team performance scores, automated communication measures were captured to obtain signal detection and duration to help understand how team communications were shared and in turn contributed to SA and performance. In addition to the communication measures, CITIES also integrated heartrate monitoring as a physiological measure of workload. All in all, the CITIES simulation was ahead of its time and the architecture utilized while being ad hoc in nature was adept at accomplishing the purposes inherent in research quests.

Some of the major results obtained through the use of CITIES included (1) showing the impact and effect of having an expert system as a team member, (2) showing how different forms of electronic-mediated technologies could impact team situation awareness and team performance, (3) demonstrating that higher team SA does not necessarily result in having higher team performance, and (4) utilizing early forms of physiological and communication measures within the context of teamwork and how this could provide useful information. Although the theories underlying some of the design of CITIES have decreased in prominence, the basic bones and premises of having a dynamic, temporal-based team simulation have withstood the test of time. As will be communicated in the next chapter, CITIES became the foundation for NeoCITIES, which continues in use today.

Jasper/Repsaj

The next simulation, termed Jasper/Repsaj, to be reviewed was actually more of what is termed a "problem set" which was housed within a simulation shell for the purpose of team research. The Jasper series was obtained through the first author's dissertation work with the Vanderbilt Learning Technology Center (through PhD advisor Dr. John Bransford) in the late 1980s and early 1990s, specifically taking place at the Air Force Research Laboratory. The use of Jasper Series represented a somewhat dramatic turn in focus in terms of the demands of the task and the kind of task that team members were required to solve. Qualitatively, Jasper provided an alternative look at team cognition, one that required deeper levels of cognition than TRAP or CITIES but was also highly immersed in perception and context.

The Jasper series is a specific problem set officially called "The Adventures of Jasper Woodbury" (The Cognition and Technology Group at Vanderbilt, 1992) and was designed to engage people in understanding, learning, and practicing "distance = rate × time" physics problems within a complex and interconnected context. The context involved crisis management and the purpose was to have a simulation representative of problem-based learning. In particular, the use of Jasper was designed to assess the degree to which a team can learn together in a realistic context and determine how memory and transfer of learning is encoded when given an individual near term transfer problem. As such Repsaj (Jasper in reverse) was a near-term analogical problem given to individuals after they had solved Jasper in the team context. Therein, the overall purpose of Jasper/Repsaj was to simulate team problem solving in a highly situated context and determine to what extent collective induction produced positive impacts on the transfer of learning and memory of an individual (see McNeese, 2000).

In this case, Jasper had all the real-world constraints and solutions for multi-step planning/problem solving embedded in a video. The video contained all the information needed to create an optimal solution path, however the problem space was ill-defined and ill-structured. Therein, participants had to determine what the actual problem was and integrate elements in context to determine how a solution path might proceed. To make things more difficult, Jasper could be solved in a variety of ways, so participants had to figure out the best solution to the challenge by contrasting and comparing different possible solutions to see which one was best. The problem-solution paths contained aspects of "distance = rate × time" physics problems, but also contained other information interdependencies that constrained possible solutions.

Specifically, Jasper required participants to figure out how to save an endangered species (an eagle) who was shot deep within a forested area without much access. This set up an urgent temporal component to the problem because the faster the eagle obtained veterinary help the higher was its probability of survival. The problem context was heavily embedded with differing means of transportation (cars, walking on foot, use of an ultralight air vehicle) and other constraints that impacted the major variable of interest: the time it took to rescue the eagle. The primary source of problem solving involved calculating D = R × T equations for various scenarios wherein secondary issues would impact solutions space (for example, the combined weight of a flyer, gas tank, and gas could be one constraint that might prohibit the use of the ultralight). The problem demanded dynamic decision making to figure out the most optimal solution. Many different solutions could be utilized, but there was one optimal solution based on insight, analytics, and planning. The problems, while being ill-defined in nature, were all presented in a highly real-world context generated by the video. The demands inherently required understanding of different roles, basic physics, and being able to test the validity of generated plans.

Jasper/Repsaj is similar to CITIES only from the standpoint they both utilize the emergency crisis management context, albeit different ones. One very unique element of Jasper is that it provides a broad exposure to problem finding and problem solving. Because the problem set contains many ill-defined elements (actually requiring a participant to parse the Jasper video into highly specific but interrelated sub-problems) it requires (using McGrath's circumplex framework) planning, creativity, and intellective decision making to generate the best solution. The way our research group utilized it for team performance was by allowing dyads (two-person teams) to work in a way that required mutual cooperative learning. This is a type of open-ended teamwork that is used for differing contexts that do not predefine how teams have to work together, so it is very valuable for real world, on-the-fly decision making that may have wicked problems embedded. Our instructions for the joint problem solving stated that the dyad was to work together and solve the challenge problem.

One of the elements in the Jasper experiments that was studied was the extent to which collective induction (Laughlin, 1999) developed and in turn how this helped to create memory and learning. The determination of how much they actually worked together (collective induction) was determined through analyses of actual videos of teamwork that were encoded along a number of dimensions (see McNeese, 2000). After completing the cooperative work condition, each individual was provided with the Repsaj problem to test whether collective induction facilitated spontaneous access of knowledge, memory response, and analogical transfer. Repsaj was a near term analogy also involving emergency crisis management but in a context of a military officer rescuing another officer experiencing frostbite in a remote area of Canada using a flying snowmobile contraption.

In contrast to TRAP and CITIES, Jasper/Repsaj tapped into a different team cognitive skill set owing to the demands presented. It provided a more naturalistic open-ended scenario for team cognition to develop. In addition to the level of the solution generated by dyads, the actual diagnosis and problem solving accomplished was evaluated through the use of a planning net (Goldman, Vye, Williams, Rewey, & Hmelo, 1992) to determine which sub-problems were actually solved effectively. The simulation represented a socio-cognitive science approach and relied on both quantitative and qualitative methods of evaluation.

During the late 1980s to early 1990s the setups for simulations were not sophisticated and often took some improvisation to create emulation of the processes needed. In the case of Jasper/Repsaj this was certainly true as the computing architecture was rather basic. The architecture/setup for Jasper was centered around a Macintosh computer connected with a laserdisc player, and a color monitor. A time-signal apparatus was used to record specific timing behavior of the participants. Dyads were required to solve the Jasper problem in a "think aloud" paradigm. Therein, three video cameras with integrated microphone systems were connected to a VCR to record the video/audio problem-solving components of a dyad's think aloud protocols. One affordance presented by a laser disc player was that it enabled participants to easily return to specific scenes in the video to replay them. This provided "real" reenactments of different subproblem elements embedded within the video case. This allowed for the process of perceptual differentiation to occur wherein contrasts and comparisons of facts, scenes, and transitions among scenes were available and could potentially be integrated. The architecture provided capture of all forms of problem-solving behavior. For the transfer problem solved by individuals, separate rooms were provided wherein the Repsaj problem was presented as verbal analogue of Jasper. This architecture achieved the desired purposes of the experiment but was inherently different from TRAP and CITIES simulations.

This particular orchestration of Jasper/Repsaj enabled experimenters to look at three phases of team cognition vis-à-vis dyadic team problem solving: knowledge acquisition and solving of a problem, transfer of learning to a near-term analogy, and memory recall of both Jasper and Repsaj problems. Among the many findings was that the transfer of knowledge is difficult for individuals when solving Repsaj as a verbal problem, even though they acquired knowledge in Jasper, a video that emphasized perceptual learning. The setup provided a lot of instruction about how to do an extended, complex experimental session that would test and evaluate dyads engaged in ill-defined, dynamic problem solving within a realistic, perceptual environment while accessing multiple measures of team and individual performance.

SUMMARY AND CONCLUDING REMARKS

Much of the work at Wright-Patterson Air Force Base was entrenched in actual applied settings such as aircraft crews, command and control teams, and engineering design teams to address issues of great concern for the Department of Defense. The history of this work set the stage for more contemporary progressions of distributed team cognition (as portrayed in the next chapter). The foundational work presented in this chapter provided much insight and ingenuity. Of course, along the way there were failures, mistakes, and false starts. This is the nature of real-world work and research where limited understanding created gaps that produce consequences. Yet much was learned and adapted for future work that helped to inspire and provide needed feedback for learning to take place. This is how lessons learned are formed and contribute to progress. As the senior author began research on teams in the early 1980s, major theoretical positions and reviews of the literature were provided by Dyer (1984), Hackman and Morris (1975), Hill (1982), and Roby (1968). Today these perspectives, reviews, and expertise have been replaced by the likes of Salas, Cooke, Mathieu, Fiore, Mohammed, DeChurch, and others as paragons of distributed team cognition. As the reader will see, many of these scientists are represented within this handbook and provide great depth of knowledge and perspicacious advice within different research areas of distributed team cognition. Likewise, there have been numerable advances in methods and measures that have facilitated advancements and greater comprehension about the target research area. Technology has deepened with complexity but has also created innovative advances that allow precise support of teamwork especially as it encompasses distributed information, places, and people. Tools have been designed that employ new outlays of data collection and data interpretation that were not thought possible 30 years ago. When one considers all these changes over the last 30 years, it is absolutely incredulous what has come to pass in advancing the state of the art of the field. This chapter has captured historical and conceptual developments from the greater good, and the next chapter will show how simulation has evolved in many different ways, and what is still possible.

NOTES

1. Ironically later in his research life at the Air Force Research Laboratory, Wright-Patterson AFB, OH he was able to become the Director of the Collaborative Design Technology Laboratory which provided research on how computer-supported cooperative work and technology could be adapted and developed for engineering design teams. This represents a salient progression from actual design work with an engineering team to study of distributed team cognition within engineering teams, therein providing an important ecological niche to much of my research that focused on supporting teams with technological innovation.
2. The first author had the pleasure to meet and interact with Dr. Warrick, one of the early pioneers of human factors, while at the Fitts Human Engineering Division. Dr. Warrick continued to volunteer at the laboratory into his 80s.
3. The use of the term "human factors" provides a generic term meant to collectively refer to human factors engineering, user-centered design, human-computer interaction, cognitive systems engineering, and human-system integration.

REFERENCES

Alluisi, E. A., Chiles, W. D., Hall, T. J., & Hawkes, G. R. (1963). *Human group performance during confinement.* USAF AMRL Report No. TDR-63-87. Wright-Patterson Air Force Base, OH: Armstrong Aerospace Medical Research Laboratory.

Alluisi, E. A. (1994). Roots and rooters. In H. L. Taylor, (Ed.), *Division 21 members who made distinguished contributions in engineering psychology.* (pp 4–22). Washington, DC: APA.

Baron, J. (2004). Normative models of judgment and decision making. In D. J. Koehler & N. Harvey (Eds.), *Blackwell handbook of judgment and decision making* (pp. 19–36). London: Blackwell.

Bransford, J., Sherwood, R., Hasselbring, T., Kinzer, C., & Williams, S. (1990). Anchored instruction: Why we need it and how technology can help. In D. Nix & R. Spiro (Eds.), *Cognition, education, and multimedia: Exploring ideas in high technology* (pp. 115–141). Hillsdale, NJ: Lawrence Erlbaum.

Brown, C. E., & Leupp, D. G. (1985). *Team performance with large and small screen displays.* AAMRL-TR-85-033. Wright-Patterson Air Force Base, OH: Armstrong Aerospace Medical Research Laboratory.

Brown, J. S., Collins, A., & Duguid, P. (1989). Situated cognition and the culture of learning. *Educational Researcher, 18*(1), 32–41.

Cannon-Bowers, J. A., & Salas, E. (1998). Individual and team decision making under stress: Theoretical underpinnings. In J. A. Cannon-Bowers & E. Salas (Eds.), *Making decisions under stress: Implications for individual and team training* (pp. 17–38). Washington, DC: American Psychological Association.

Chapanis, A. R., Garner, W. R., & Morgan, C. T. (1949). *Applied experimental psychology.* New York: John Wiley & Sons.

Clapper, T. C. (2018). *Facts and fears: Hard truths from a life in intelligence.* New York: Viking.

Cognition and Technology Group at Vanderbilt. (1992). The Jasper experiment: An exploration of issues in learning and instructional design. *Educational Technology Research and Development, 40*(1), 65–80.

Cooke, N. J., Gorman, J. C., Duran, J. L., & Taylor, A. R. (2007). Team cognition in experienced command-and-control teams. *Journal of Experimental Psychology: Applied, Special Issue on Capturing Expertise across Domains, 13,* 146–157.

Cooke, N. J., Rivera, K., Shope, S. M., & Caukwell, S. (1999). A synthetic task environment for team cognition research. *Proceedings of the Human Factors and Ergonomics Society 43rd Annual Meeting* (pp 303–307). Santa Monica, CA: Human Factors and Ergonomics Society.

Dyer, W. G. (1984). *Strategies for managing change.* Reading, MA: Addison Wesley Publishing.

Eimer, E. O. (1987). *When decision aids fail.* AAMRL-TR-87-035. Wright-Patterson Air Force Base, OH: Armstrong Aerospace Medical Research Laboratory.

Endsley, M. R. (1995). Measurement of situation awareness in dynamic systems. *Human Factors, 37*(1), 65–84.

Festinger, L. (1962). Cognitive dissonance. *Scientific American, 207*(4), 93–107.

Gardner, H. (1985). *The mind's new science: A history of the cognitive revolution.* New York: Basic Books.

Goldman, S. R., Vye, N. J., Williams, S. M., Rewey, K., & Hmelo, C. (1992, April). *Planning net representations and analyses of complex problem solving.* Paper presented at the annual meeting of the American Educational Research Association, San Francisco, CA.

Grether, W. F. (1995). Human engineering: The first 40 years 1945–1984. In R. J. Green, H. C. Self, & T. S. Ellifritt (Eds.), *50 years of human engineering: History and cumulative bibliography of the Fitts Human Engineering Division.* Wright-Patterson Air Force Base, OH: Crew Systems Directorate, Armstrong Laboratory, Air Force Materiel Command.

Hackman, J. B., & Morris, C. G. (1975). Group tasks, group interaction process, and group performance effectiveness: A review and proposed integration. In L. Berkowitz (Ed.), *Advances in experimental social psychology* (Vol. 8). New York: Academic Press.

Hill, G. W. (1982). Group versus individual performance: Are n+1 heads better than one? *Psychological Bulletin, 91*(3), 517–539.

Hinsz, V. B., Tindale, R. S., & Vollrath, D. A. (1997). The emerging conceptualization of groups as information processors. *Psychological Bulletin, 121*, 43–64.

Hollins, J., Hutchins, E., & Kirsh, D. (2000). Distributed cognition: Towards a new foundation of HCI. *ACM Transactions on Computer Human Interaction, 7*(2), 174–196.

Johnson, D.W. & Johnson, R.T. (1994). *Learning together and alone* (4th ed.), Needham Heights, MA: Allyn and Bacon.

Kidd, J. S. (1961). A comparison of one-, two-, and three-man work units under various conditions of work load. *Journal of Applied Psychology, 45*(3), 195–200.

Kincade, R. G., & Kidd, J. S. (1958). *The effect of team size and' intermember communication on decision-making performance.* WADC TR 58-474. Wright-Patterson Air Force Base, OH: Aero Medical Laboratory, Wright Air Development Center.

Klein, G. A., Orasanu, J., Calderwood, R., & Zsambok, C. E. (Eds.). (1993). *Decision making in action: Models and methods.* Westport, CT: Ablex Publishing.

Laughlin, P. (1999). Collective induction: Twelve postulates. *Organizational Behavior and Human Decision Processes, 80*(1), 50–69.

Letsky, M. P., Warner, N. M., Fiore, S. M., & Smith, C. A. P. (Eds.). (2008). *Macrocognition in teams.* Burlington, VT: Ashgate.

Lindsay, P. H., & Norman, D. A. (1977). *Human information processing: An introduction to psychology.* New York: Academic Press.

Lorge, I., Fox, D., Davitz, J., & Brenner, M. (1958). A survey of studies contrasting the quality of group performance and individual performance, 1920–1957. *Psychological Bulletin, 55*(6), 337–372.

Martin, E., Lyon, D. R., & Schreiber, B. T. (1998). Designing synthetic tasks for human factors research: An application to uninhabited air vehicles. In *Proceedings of the Human Factors and Ergonomics Society annual meeting* (pp. 123–127). Santa Monica, CA: Human Factors and Ergonomics Society.

McBride, D. J., & Brown, C. E. (1989). Team performance in a dynamic resource allocation task: The importance of heuristics. In *Proceedings of the Human Factors and Ergonomics Society 33rd annual meeting* (pp 831–835). Santa Monica, CA: Human Factors and Ergonomics Society.

McCormick, E. J. (1957). *Human engineering.* New York: McGraw-Hill Publishers.

McGrath, J. E. (1984). *Groups: Interaction and performance.* Englewood Cliffs, NJ: Prentice Hall.

McNeese, M. D. (1992). *Analogical transfer in situated cooperative learning.* Unpublished doctoral dissertation, Vanderbilt University, Nashville, TN.

McNeese, M. D. (2000). Socio-cognitive factors in the acquisition and transfer of knowledge. *Cognition, Technology, and Work, 2*, 164–177.

McNeese, M. D., & Brown, C. E. (1986). *Large group displays and team performance: An evaluation and projection of guidelines, research, and technologies.* AAMRL-TR-86-035. Wright-Patterson Air Force Base, OH: Armstrong Aerospace Medical Research Laboratory.

McNeese, M. D., & Brown, C. E. (1986). *Large group displays and team performance: An evaluation and projection of guidelines, research, and technologies.* AAMRL-TR-86-035. Armstrong Aerospace Medical Research Laboratory, Wright-Patterson Air Force Base, OH.

Meister, D. (1999). *The history of human factors and ergonomics.* Mahwah, NJ: Lawrence. Erlbaum Associates.

Mesmer-Magnus, J., & DeChurch, L. (2009). Information sharing and team performance. A meta-analysis. *Journal of Applied Psychology, 94*(2), 535–546.

Morrissette, J. O. (1958). An experimental study of the theory of structural balance. *Human Relations, 11*(3), 239–254.

Myers, B. A. (1998). A brief history of human computer interaction technology. *ACM Interactions, 5*(2), 44–54.

Nardi, B., and Miller, J. (1991). Twinkling lights and nested loops: Distributed problem solving and spreadsheet development. *International Journal of Man-Machine Studies* 34:161–184.

Neisser, U. (1967). *Cognitive psychology.* New York: Appleton-Century-Crofts.

Neisser, U. (1976). *Cognition and reality: Principles and implications of cognitive psychology.* New York: Freeman.

Pattipati, K. R., Kleinman, D. L., & Ephrath, A. R. (1983). A dynamic decision model of human task selection performance. *IEEE Transactions on Systems, Man, & Cybernetics, SMC- 13*(2), 145–166.

Pea, R. D. (1993). Practices of distributed intelligence and designs for education. In G. Salomon (Ed.), *Distributed cognitions* (pp. 47–87). New York: Cambridge University Press.

Rasmussen, J., Pejtersen, A. M., & Goodstein, L. P. (1994). *Cognitive systems engineering.* New York: Wiley.

Roby, T. B. (1968). *Small group performance.* Chicago, IL: Rand McNally.

Salas, E., Fiore, S. M., & Letsky, M. P. (Eds.). (2012). *Theories of team cognition: Cross-disciplinary perspectives.* New York: Taylor & Francis Group.

Tversky, A., & Kahneman, D. (1974). Judgments under uncertainty: Heuristics and biases. *Science, 185*(4157), 1124–1131.

Vygotsky, L. S. (1978). *Mind in society: The development of higher psychological processes.* Cambridge, MA: Harvard University Press.

Wellens, A. R. (1990). *Assessing multi-person and person-machine distributed decision making using an extended psychological distancing model.* AAMRL-TR-90-006. Wright-Patterson Air Force Base, OH: Armstrong Aerospace Medical Research Laboratory.

Wellens, A. R. (1993). Group situation awareness and distributed decision making: From military to civilian applications. In J. Castellan (Ed.), *Individual and group decision making: Current issues* (pp. 267–291). Hillsdale, NJ: Lawrence Erlbaum Associates.

Williges, R. C., Johnston, W. A., & Briggs, G. E. (1966). Role of verbal communication in teamwork. *Journal of Applied Psychology, 50*(6), 473–478.

Wilson, D., McNeese, M. D., & Brown, C. E. (1987). Team performance of a dynamic resource allocation task: Comparison of shared versus isolated work setting. In *Proceedings of the 31st annual meeting of the Human Factors Society* (Vol. 2, pp. 1345–1349). Santa Monica, CA: Human Factors Society.

Wohl, J. G. (1981). Force management decision requirements for air force tactical command and control. *IEEE Transactions on Systems, Man, and Cybernetics, 11*(9), 618–639.

3 Reflections on Team Simulations—Part II
Contemporary Progressions

Michael D. McNeese, Nathaniel J. McNeese, Lisa A. Delise, Joan R. Rentsch, and Clifford E. Brown

CONTENTS

INTRODUCTION

One purpose of the *Handbook of Distributed Team Cognition* is to examine the crossroads of looking back and looking ahead while reviewing contemporary research and practice in how teams address currently complex and challenging problems. Part of these crossroads is the contributions of team simulations in terms of concepts, value, and worth to the scientific community. Hence the previous chapter and this chapter provide a holistic view—from a particular reflection—into how team simulation has transpired and worked through the confines of a given research group (more recently referred to as the MINDS Group).[1]

Consideration of teamwork in contemporary culture may produce different meanings than what was presented in the previous chapter, which focused more on historical research practices. In today's world teams are very much prevalent in many kinds of work domains and fields of practice including medical and health, transportation, logistics, maritime operations, information and cybersecurity, offshore operations, manufacturing and production, and aviation, to name a few, and very much in evidence in a variety of organizations, businesses, government, and industry. Although this is not necessarily different from the previous practice of teamwork, there are some distinct changes in teamwork that are notable that lead to a current understanding of distributed team cognition, and why research has evolved given some of these differences. There are two areas that especially stand out as contributors to change in teamwork: (1) technological innovation and (2) contextual complexity.

Technological Innovations Underlying Teamwork

Technological innovation is rampant across contemporary society wherein computational power and digital capability is omnipresent. Information technology developments have coupled power and capability to produce new socio-media tools, cognitive architecture platforms, group interfaces, and collaborative support applications that directly facilitate how people think and act within teams. Many of these innovations are now provided directly through the internet or through the use of smartphones where apps can be downloaded and quickly used for real work considerations. Computers are allowing a greater diversity of teams to come together from great distances all across the world with much ease. As pointed out in our previous chapter, historically, many traditional teams were limited in that they were typically only in the same place at the same time. The ability to distribute teams at remote distances and have synchronous and asynchronous interaction was limited (primarily to use of phones). The ability to fully integrate members at a distance at the same time was pretty much unavailable. This created intellectual barriers as diverse and multiple perspectives on team problem solving and decision making were subject to who "was in the given context at the time." Not only was distributed work limited but distributed information was usually not available to teams for use in their work. Information such as text, graphics, video and photographic feeds, and other forms of media and data were limited to just what was available in the local context. As mentioned in the last chapter, research focused on team cognition wherein information and collaborative technology was predicted for the future but not really developed to any extent.

Insightful researchers such as A. Rodney Wellens (1993) simulated team decision making at distance for purposes of comparing different forms of electronic media (text-based messages in computer setups, telephony, and cable-generated video to simulate video teleconferencing). This projected what the future would become 30 years later, which is where we are today, except the innovation of today is far beyond the expectations of 30 years ago. The power of the internet alone and what it has meant in terms of distributed teamwork and distributed information access has been a tremendous boost in changing how people work together. The presence of such apps as Facebook, Twitter, and Instagram have changed the nature of communication, collaboration, coordination, and perception in unique ways. Therein, whereas in the past much of the research focused on *team* cognition, the research of today is clearly resonant with *distributed* team cognition. In turn, team simulators must adapt and be representative of real-world technological changes in today's society rather than just show ad hoc add-ons.

In addition to technology appropriating instant collaboration via digital software applications (e.g., Slack, Zoom, Google Docs) the processes of teamwork/cooperative work over the years have been inspired through various theories and approaches resulting in the design and use of various tools/supports (see Schmidt & Bannon, 2013 for review). Whereas in the past these computing technologies, appropriated for team use (e.g., team decision aids), were awkward for the user, today they are seamless, integrative, and have been designed from more user-centric and awareness perspectives (Gross, Stary, & Totter, 2005; Tenenberg, Roth, & Socha, 2016). Our first foray into understanding computer-supported cooperative work emphasized the various approaches that could be taken and how the resultant designs of groupware where representative of those approaches, but in particular how a technology-only perspective was wholly inadequate (McNeese, Zaff, & Brown, 1992). Since then the role of affect, emotion, and beliefs (Hudlicka, 2003), context and field of practice, and user experience have become much more paramount in the design of systems.

Additionally, the collaborative technologies of today emphasize an important concept within distributed cognition—information sharing—whereby distributed teams can not only access reams of information but they can work through distributed shared objects, interactive group interfaces (McNeese, Theodorou, Ferzandi, Jefferson, & Ge, 2002), all with built-in documented updates and traces of who contributed what when and at what time. This creates distributed cognition with high temporal flow present. Many perspectives, much diversity, and multiple levels of intelligence can be easily rendered within these kinds of collaborative suites. However, that is not the total picture, as distributed cognition also encapsulates new formulations of perception combined with information sharing through the use of advanced technologies which have been referred to as *virtual presence*. Being able to perceive situations remotely is a way to enhance individual and team situation awareness across multiple events. A simple application of this is the vast network of cameras that one can now access online to "see" situations remotely.

During the 2013 Boston Marathon bombing distributed information sharing was of utmost importance to track situations and perform search and rescue activities. This provides a perceptual anchor for communications and joint coordination of activity. The networks of cameras provided source material for use of face recognition technologies which also facilitated identification and tracking of subjects. This is just a simple example but shows how the power of many different views from remote locales can contribute to sensemaking, surveillance, and problem solving. The obvious use of this is in the area of police cognition to "see" a situation unfolding from different temporal and visuospatial perspectives. However, it should be noted that the distributed cognition in this kind of event can also go wrong. Although crowd sourcing of faces and cogent pictures were actively being distributed across sites on the internet (e.g., Reddit), this produced numerous false alerts which can then lead police down the wrong rabbit holes. This is example where distributed cognition was active and engaged but ironically it did not necessarily lead to all the correct outputs. There were also ethical issues at play as people were falsely accused in the haste of trying to identify subjects. This signals that the collective can be more problematic than a simple team and that one must be careful given results obtained. The work our group did in the mid-1980s on information sharing using small and large group displays (McNeese & Brown, 1986) has certainly come a long way in more than 30 years but the premises underlying information sharing, interdependent cognitive processes, and team situation awareness are still intact.

The world of virtual environments has skyrocketed providing teams with a unique interface that can bring people together even though they are remotely located. Virtual environments are created by using multimodal, high bandwidth technologies (such as 3-D interactive graphics, 3-D sound space, interactive touch with virtual controls, and dynamic integration of sensory perception). One of the most rewarding experiences the senior author had during the 1980s at the USAF Harry G. Armstrong Medical Research Laboratory was flying the Super Cockpit that Dr. Thomas Furness had designed and operationalized. Dr. Furness is widely known as the "grandfather of virtual and augmented reality" (Clapway, n.d.). The Super Cockpit was basically an entire virtual cockpit inside of a very large helmet, which was wired with sensors for information flow and interactive touch-feedback. The immersive experience emulated being inside an advanced fighter cockpit. This was a unique kind of simulation as one could use the apparatus to fly and perform a mission, at least in experimental fashion. Technically, the Super Cockpit was originally designed for a single-pilot immersion, however concepts were hypothesized wherein multiple people could engage in a virtual world. While that was primarily a vision of the 1980s, virtual environments are now feasible for teamwork and have been used commercially through such applications as the Second Life® virtual world. Virtual environments and augmented tools have since taken off with new commercially available technologies (e.g., Oculus Rift, Google Glass, HTC Vive) which are predicated on Furness' early developments and experiments. Virtual and augmented reality are one example of where digital information technology intersects team processes producing distributed team cognition.

The software application of a decade ago termed Second Life is an example of where brilliant 3-D virtual environments could create a medium for team interaction through the use of avatars representing individuals within a constructed world. The avatar that represents "you" can communicate with other avatars in the virtual community, take joint actions based on the affordances that can be picked up in the world, and just "hang out" and engage in social interaction with others. The world also provided certain commercial services embedded in the world, which utilized online financial economies. This unique collaborative technology provided the means for a community of actors to engage in distributed team cognition. The environment (graphics, auditory space, and information services) could be constructed according to the purpose of the interaction. For example, if one was desiring to create a Second Life virtual course that studied ancient Rome, the interface could be designed to emulate the ancient ruins with specific places where avatars could meet and interact and discuss various elements of culture, philosophy, and architecture. Each student would have an avatar that represented them in the class. Although this sounds futuristic, experiments in Second Life have looked at how this kind of environment facilitates team problem solving (see McNeese, Pfaff, Santoro, & McNeese, 2008), therein testing it as a means of producing new forms of teamwork. The Second Life virtual world was intriguing and it required monthly payments to subscribe to it (or to "rent an island" in the vernacular). However, sustaining it through the economic model and figuring out exactly what advantages it had for more traditional team environments has been challenging. Therefore, it did not exactly catch on as a research tool.

Contextual Perturbations of Teamwork

Technology has advanced distributed team cognition research in creative and prodigious ways, but teamwork enacted within the contemporary culture also has been interpreted more from a contextual-situated lens. Put simply, the role and influence of the environment has become increasingly salient in understanding individual and team activities. The predecessor book that lent credence to this handbook, *New Trends in Cooperative Activities* (McNeese, Salas, & Endsley, 2001), represents a kind of juncture between traditional work in team cognition and a newer vision of cognitive-social-contextual interactions. This newer view has been referred to as situated cognition (Brown, Collins, & Duguid, 1989; Greeno, 1998; Young & McNeese, 1995) or distributed cognition (Hollan, Hutchins, & Kirsh, 2000; Hutchins, 1995; Salomon, 1993) in that cognition is always in reference to and situated within a specific place, a specific context, and a given environment. This could be inclusive of physical characteristics of a place, the social milieu that surrounds and leads to team cognition, the cultural backdrop that makes sense out of actions, and the technology that supports cognition. Some researchers use situated and distributed cognition interchangeably, but it has been our experience that situated cognition often refers to tasks, events, episodes, and situations in which humans are coupled with action and embedded in the social milieu that affords intentionality through action. Distributed cognition highlights social activities as being highly distributed and represented with information, objects, space, and people. Although these differences are subtle, they can be meaningful to certain research theories and levels of understanding.

Recent views emphasize the predominance of ecological psychology through concepts such as affordances the environment supplies that agents act upon (always in relation to one another) and the idea of direct perception where designs (Norman, 2013) make it likely that information is directly picked up from the environment an agent is acting upon (Gibson, 1979). Ecological approaches emphasize the availability of the dynamic environment where movement through space and time are important for experiencing situations and problems, whether it be at the individual or the group level. Dynamics like these can be specified by information distributed across a context and lead to enhanced awareness of activity and situations occurring in that environment. Designs for team simulations started to utilize Gibsonian concepts to make the simulation more reflective of real-world behavior to the extent possible, contributing to the ecological validity, specification of salient information, and the fidelity participants should experience.

Ecological psychology and situated cognition perspectives (Norman, 2013) that place an emphasis on understanding real-world behavior (in the context it occurs) have concomitantly contributed to a change in methodological approaches in team cognition. Traditional cognitive psychology approaches typically are coupled with quantitative methods in a highly controlled experimental laboratory whereas ecological/situated approaches emphasize more qualitative research methods (e.g., ethnography, case study, contextual inquiry, scenario-based design) although they may also utilize quantitative study. Qualitative methods often place reliance upon contextual perturbations that can change the way an individual or team perceives a problem state.

Likewise, qualitative methods may capture how a team prioritizes contingencies that are emergent owing to dynamic events or dynamic movement. Qualitative methods can also reveal impacts and importance of the individual differences within the team that may go unnoticed in quantitative methods. During our use of the qualitative method, concept mapping (Zaff, McNeese, & Snyder, 1993) knowledge integration could be shown to develop between the individual and team-level performance which showed the interdependence among team members.

One aspect of knowledge that is important is how beliefs may translate/evolve from individual differences and get integrated into a team mental model and how the context/culture influences this process. Belief formation may arise through (1) interaction with others and (2) repeated experiences with specific, similar contexts. Although there are other sources for beliefs such as biases, dogmatic convictions, and influence of a trusted source, beliefs arising out of social interaction and experience may highlight the role of individual differences. It may be the case that team conflict is present in a situation as a function of different team members holding to their belief systems obtained through experience which may be oppositional to other team members' beliefs. Conflicting beliefs can be a real barrier for team cognition and prevent formation of shared mental models. Although team members show unique individual differences and cultural biases, at the same time, their combined set of team processes, requisite knowledge, and skills/abilities offer the opportunity to congeal effective, shared mental models, apropos for the situation. Qualitative methods such as ethnography or concept mapping can provide, capture, and analyze the processes underlying how individuals put their thoughts together to formulate team cognition (or not, in the case of when teams fail to reach common ground). Convergence can be motivated by the need for interdependence to solve a problem at hand (e.g., when different roles on the team produce strands of information that need to be tied together at the right point in time to act on a problem).

The qualitative approach may also help reveal the basis for how cultural variation imprints on individuals and small team behavior. With the global economy being the norm for businesses and for military actions being initiated across nations and governments, team composition will continue to be predicated on determining cultural meanings across team members. For recent research on this refer to Endsley (2016).

Because technology has enabled team participants to be present in different team orchestrations—either at the same time or at different times—they may be required to switch among the contexts that are appropriate to the team to which they are contributing. Context switching is indicative of complex cognition-to-context awareness and is a more difficult task than just having to keep track of one context. Later in the chapter, the use of this type of context switching arrangement is elaborated in one of the simulations built to study distributed team cognition in army command and control tasks (Fan, McNeese, Sun, Hanratty, Allender, & Yen, 2009).

The senior author has always been sensitive to ecological and environmental influences on cognition, because the primary approach in contemporary team research resonates around situated cognition, naturalistic decision making (Klein, Orasanu, Calderwood, & Zsambok, 1993), and real-world constraints on emergent problems (McNeese & Forster, 2017).

CURRENT EXAMPLES OF DISTRIBUTED TEAM COGNITION

As a review of contemporary distributed team cognition is distilled in this chapter within the purview of team simulation, it is useful to consider three real-world examples of how team cognition is highly interdependent on distributed information, contextual perturbation, spatial-temporal integration, and the help/support afforded by information and collaborative technologies. The examples focus on emergency crisis management, uninhabited air vehicle operation, and practice of remote medicine on the battlefield.

The first example revolves around an actual crisis event the senior author experienced at an engineering organization within the confines of a military air force base. Because the event was terrifying and dramatic there is much familiarity and recalled memory associated with it. The focus for this first example then is emergency crisis management which is also the context that inspired our current NeoCITIES simulation. It is a somewhat limited example because it occurred in the late 1970s when information and collaboration technologies for the most part were not yet present in most organizational environments. However, it provides meaning for contextual variability, sensemaking, and emergent perception of a situation as it unfolds within a collective, distributed environment. And it is instructive to see by contrast how the presence of distributed team cognition would have been different if the event had happened in today's culture.

Emergency Crisis Management

There is a lot of teamwork that develops and is present at the immediate vicinity surrounding a crisis event. Likewise, multiple distributed teams also are drawn into action with additional resources to be managed (therein a team of teams context is in effect). In crisis events, the information that is known versus unknown often dictates the coordination and appropriateness of a team response, and how that response will be worked out given the particulars of the context. As changes happen more context perturbations will need to be considered and factored into the response. The elements of the crisis that are difficult and challenging then are the level of uncertainty that exists, the degree and level of risk that is impending and therein the necessity for quick, immediate action, and the varying degrees of stress, turmoil, and chaos that have to be managed to the extent possible. When these elements are all present action needs to be taken immediately to mitigate risk but at the same time safety is of utmost concern. The elements must be thoughtfully considered with the constraint of safety as a foremost value and concern for all involved. The scale of crises can range from mild (concerned and involved) to the more extreme (volatile and dangerous). A trashcan fire at a business might be an example of a mild crisis. In this case it is pretty well-defined and the actions to mitigate risk are straightforward. However, if the action is not taken immediately then the event can escalate into to a full-fledged crisis. On the other hand, an example of an extreme crisis might be a mass shooting wherein the situation is volatile and dangerous for many. When a shooting occurs, chaos often ensues and is evident by a lack of understanding of what is happening. Information distributed across multiple sources may be hidden or partially hidden to other people in the context. In such high-risk and stressful situations people may quickly have to come together to figure out what is going on and/or to protect themselves to survive, and action steps may be of high importance to help facilitate mitigation of the risk.

The senior author of this paper was involved in an actual crisis management event which took place at a military installation in the continental United States. This event took place around 40 years ago and it is informative to consider some of the aspects inherent in distributed team cognition as it arises on the fly in the midst of a crisis emerging:

> I was working in the job as a designer for an avionics systems engineering group when this event occurred one morning. The context was a large engineering building where various individuals and teams were distributed at work on multiple different projects at various locales throughout the building. The layout of the building was such that it had a central high-bay area where actual systems could be built and tested with the proper equipment. I was at work on a drafting table/desk, deeply focused on completing the details of a drawing, when I heard a woman screaming very loudly about twenty-five feet away. This was an unusual occurrence that is experienced infrequently. This first cue indicated that a crisis might be in the making or it might be

some other kind of anomalous event taking place and immediately puts one on high alert. As I investigated what was going on several things happened. In my mind, as I recall, this was a warning of something odd happening therefore out of the ordinary, which immediately increased stress and uncertainty, a state of concern, and an awareness of my surroundings. There was also a sense of uncertainty present. Because of this I decided to investigate the source of screaming and quickly walked towards it. In essence this is an example where events are still uncertain and ill-defined, and investigation hopefully leads to more information and potential action to reconcile what might be a pending crisis.

As I moved towards the source of screaming, my investigation was disrupted by an Office of Security Investigation (OSI) agent holding a weapon. My stress skyrocketed upon seeing this. As he turned towards me (with the pistol pointed directly at me) he asked in an urgent voice, "Who are you and what are you doing here?" I identified myself as a government contractor—my office was in the back—and I came to see "what was going on." His response still leaves me with a chill 40-some-odd years later, "Your boss was just shot to death upon leaving this building a few minutes ago—the screaming you heard is his girlfriend in the office just adjacent to yours." My boss was a military officer. My fear and anxiety significantly increased even further to the point of shock upon hearing his next sentence, "You had better secure yourself and prepare for the worst, this may be a terrorist event—we are not sure what is going on—but be prepared for gunfire, bombs, and, if need be, find an exit in any way possible." Now my heart was pounding and at the same time I was trying hard to think and figure out what to do. A perfect example of how emotions shutdown attention and memory—actually an odd form of attention deficit disorder.

The OSI agent at this point also commanded me to go back to my work area rather than venturing into the high-bay area; I quickly conformed. The high-bay was primarily surrounded by offices and a few labs but it was a very large open space which spanned upwards probably three floors. This area was literally in the process of being secured. As soon as he said to secure yourself in your work area I immediately observed what looked like SWAT teams coming into the high-bay with automatic weapons drawn, German police bomb dogs, and general chaos ensuing as employees where scattering and running into various offices (emotion and stress were increasing across the board).

I had decided if I heard gunfire, I would slam my heavy office chair through the window in my office and run out of the back of the building, assuming the gunfire was not coming from that direction. This was my impromptu escape plan given the information I had. As I look back on this situation the information was rather limited, other sources of information were distributed across the building and were not available—they could not have been seen or heard. This timeframe was before computer connectivity to the internet, before cell phones existed, and old-style telephones were the main form of remote communication but I think the lines were jammed so they were ineffective. Direct communication with others on other teams was nonexistent unless people were together within a given office or lab space. As military police started to swarm upon the building and determine if it was secure or to determine what was occurring, they communicated on their headsets but the rest of us were in the dark. Police I believe were now communicating to the extent possible with different offices/lab spaces throughout the context while also investigating/securing these spaces. This is an example of chaos and fear in the midst of very little communication. Eerily, I could hear many people screaming now, sobbing and crying, after being told of what had happened. The officer shot was well-liked by many and this created a monumental emotional layer to the context that was emerging.

> Command was taken over by military police as they searched the building for potential terrorists, bombs, or other clues. I could hear loud speakers in the high-bay but the sound was muffled so I could not discern what was being said. The building was under lockdown until the middle of the afternoon. However, after about 2.5 hours the military-in-charge were certain that it was not a terrorist event and communicated by word of mouth that people could move freely around the building. This was the first sense of relief and psychological release for me and many others. Prior to this point many of us thought this might be the end. This is when we found out the real story of "what was happening" and when collective awareness increased as employees in different offices began migrating to share or to seek the help of their friends or bosses in other parts of the building, and any additional information that could be found out. This is when communication started to ensue. Of course, this is the time and setting where rumors run rampant—and they did. As the day went on I found out that my boss had been stalked for several weeks by a woman who was his former girlfriend. The woman screaming was his current girlfriend. Apparently, his former girlfriend had a mental disorder and unfortunately had a psychotic episode, killing him on his last day of duty as he left the engineering building. I was the last person to talk to him as I wished him well in his next duty station. Haunting.

If this emergency crisis would have occurred in today's world, information and communication technologies would have enabled much more sharing of distributed information through smartphone communication and capture of different locales throughout the building (and outside the building). Collaborative picture sharing using different apps on smartphones and the feedback from multiple cameras positioned throughout would have enabled a much greater perception of how the context was perturbed (whether anomalous events were taking place across a distributed environment). Distributed team members would have engaged in texting and actual phone conversations without phone lines going down.

In conclusion, this situation was not what it first appeared to be. The killer—after the shooting occurred—escaped by changing clothes and then going out an alternative exit at the installation and was found in hiding a few days later in a forested area near the installation. This is an example of emergency crisis management wherein the context is explored to come up with answers, information (among the police and agents) was shared to figure out the situational awareness of events, and most of the collective intelligence among the workers was very limited (not distributed). The main team orchestration was through the military police on-site who spread out across the building trying to pick up clues and evidence. The one element which was very extreme was the ensuing stress that left many sick, depressed, and with the need for counseling. In his limited capacity, the senior author of this chapter tried to provide leadership and help to others using his psychology background, but this was a terrible thing to have happen.

As researchers look to make team simulations more ecologically valid, representative of real-world problems and issues, and produce scaled worlds that emulate the cognitive demands within contexts of teamwork, there are natural limitations. As can be determined from this example, the whole arena of stress, volatility, emotional affect, and danger are difficult to create or replicate. In many instances, it would be immoral and unethical to consider inducing such states. Yet, these states contribute to the holistic nature of problems and how teams or collections of people respond and experience them. Certainly, many of the distributed information, and situation awareness, conditional changes, temporal constraints, and workload attributes are replicable with contextual realism and should be present within a scaled world simulation. But it will remain a major challenge to capture these other components. The qualitative method is imperative for assisting understanding of these kind of states and contributes to triangulating the comprehension of complex systems.

Uninhabited Air Vehicles

The second example of contemporary distributed team cognition is situated in a military context. From 1995 to 2000, many of the initial engineering designs of uninhabited air vehicles (UAVs) were transformed from ideas to realistic products for use in actual encounters. There have been many military drones engineered to remotely fly on their own for different purposes (e.g., close air support, reconnaissance, intelligence gathering, and even combat). Yet that flight is only possible if there is a competent and alert crew supporting the UAV flight, performing interleaved functions, and acting to achieve the mission objective. Hence, a more common designation of these aircraft today is "remotely piloted aircraft" because it highlights the necessity of humans and teamwork as part of their flight/mission. Stated another way UAVs represent semi-autonomous vehicles wherein they contain embedded automatic subsystems for specific purposes. However, their overall mission is controlled and overseen by a crew. Different kinds of UAVs (e.g., Predator, Reeper, Global Hawk) have achieved marked success over the last 20 years and have been a strategic asset in the war on terror in Afghanistan and Pakistan and other regions. Having mentioned the success of UAVs they also can have flaws in specific automation functions as well as human errors within the team controlling them which can result in mistakes and failures. Although this section focuses on military use of drones there have also been accidents in the commercial use of drones on various occasions.

UAVs are an example of distributed team cognition occurring within complex contexts that change and require updated system dynamics for success. First, the ability of a plane to fly remotely as controlled from another location at distance is one example of distributed control. Pilots who fly in the cockpit are now offered the capability to fly planes remotely without the threat of being shot down or of being involved in a crash in a dangerous environment. This provides a tremendous human benefit of saving lives of pilots and crews, and an economic benefit of system reuse and precision-based strikes. But other elements of distributed team cognition are part of a successful orchestration as well.

Second, distributed information associated with flying the plane itself, along with the specific functions needed for the mission of the aircraft, are being processed across many channels and propagated from multiple sources (e.g., the area of information fusion). These sources of information may need to be combined, refined, processed, and amplified for greater meaning, and then distributed to various team members to provide updates on the state of the aircraft, the threat, and the environment as the drone engages in its mission.

Third, because these kinds of information streams are voluminous they must be managed, controlled, and displayed in a manner that is human-centered for the distributed crew at a point in time when they are most relevant (the right information at the right time in the right place for the right human). In operational control, it may be the case that a crew has oversight, management, and control of multiple vehicles simultaneously which makes complexity, risk, workload—and hence human errors—increase dramatically. The crew would typically perform different roles such as external pilot, air vehicle operator, or other innate specialists (e.g., aerial photography specialist) that require interdependence and integration of effort inclusive of team situation awareness. The structure of work within the UAV system is representative of distributed team cognition as there are some tasks that individuals must do independently based on their crew role, while there are other times where they need to switch to shared team tasks (Nisser & Westin, 2006). In performing both individual and team-level tasks, the crew relies to a high degree on multiple facets of automation, satellite communications, radar, sensors, video feeds, etc. to gain information that helps fulfill the mission requirements. Given the dynamics embedded in UAV missions, the crew requires transitions between individual and team tasks, while subsystems may also require manual or automatic control. Another transition issue that complicates matters is the point when the crew needs to be replaced with new crew members due to fatigue. This state requires the necessity of handoffs, which can be problematic in team cognition (Tvaryanas, 2006).

At given points in time, the UAV crew may also require communications with ground personnel or other personnel specialists remotely located to provide additional insights, awareness, distributed workload, or spotting for targeting or contextual knowledge. Hence, the cognition and coordination components can place high demands on the crew that challenge team situation awareness and potentially cause information overload. When multiple UAVs are being controlled simultaneously, this is especially true. Models and team architectures have been proposed and studied in team simulation for helping a crew control multiple robots, such as UAVs, whereupon the idea of backup behavior was found to be critical for model success (Gao, Cummings, & Solovey, 2014).

When you add all of these components together, the UAV represents a very complex, distributed cognitive system that is reliant on human, technological, and contextual factors, all of which are highly interdependent to achieve mission objectives. It provides a contemporary example of distributed team cognition that is nonlinear and operative in layers and degrees where time, precision, and calibration are very critical for successful outcomes.

Because of the high degree of synchronicity among system functions during flight, shared knowledge among crew members, and the command, control, communication, collaboration, and intelligence functions that must all come together through human-system interaction, the UAV context is perfect for consideration of team simulation to test various aspects of distributed team cognition. Indeed, Nancy Cooke and her associates have done just that with their high-fidelity simulation of a UAV crew (Cooke & Shope, 2002), which provides a variety of configurations, experimental control, and a wide range of dependent team performance measures. As modeling and simulation have become more advanced, the role of human-autonomous systems becomes more appealing and potentially available for UAV operations. Inherently, intelligent and integrative models may replace some of the crew member functions to make the UAV system even more complex (McNeese, Demir, Cooke, & Myers, 2018), therein requiring profound insights into human-centered computing. Research studies within this area will be extremely valuable for understanding distributed team cognition.

Remote Medicine on the Battlefield

The third example also related to distributed team cognition is in a military setting. The history of warfare has produced battles in which there have been many losses of human life but left many behind on the battlefield injured without much hope. Those who were injured were assisted by trained medics who cared for them with tools of the trade (mainly medications and certain instruments which could be used appropriately until more in-depth care could be received), and through the expertise they had gained with others who were in similar predicaments. In other cases, soldiers had to make do on the fly in the best possible way. In certain situations, helicopter/air relief could be called in to lift out the injured and fly them to hospitals out of the warzone.

In today's world, the promise of technology and information availability provide an innovative way to address battleground situations that involve injury. Using telemedicine a team of medical providers can have remote access to diagnosis and even enact surgical procedures (see Reichenbach et al., 2017). Accessing physicians and specialists for interpretation, guidance, and communication about various injuries is possible wherein imagery, pictures, and radiological imprints can be shared. The ability to do robotic surgery is commonplace in hospitals now but to be able to perform surgery, say 1,000 miles away, at a remote medical installation on the battlefield involves a complex array of distributed information, communication, collaboration, coordination, and control, all integrated in real time with precision and with visual and auditory feedback.

Obviously, having appropriate team processes and personnel in place for this major endeavor is necessary as well. This is an example of distributed team simulation wherein being able to have synchronicity and a high degree of sharing across several environments is of utmost importance. It also represents a team-to-team orchestration as the medical personnel/team at the onsite battleground needs to collaborate with a remote medical group to ensure joint operations ensue with accuracy and with adroit knowledge present. This example emphasized the importance of collaborative information sharing under high-risk conditions where the consequences of a mistake or misjudgment could be catastrophic. In summary, this kind of situation is perfect for team simulations involving remote procedures to enable familiarity, practice, and dealing with emergency events at specific focal points in the individual-to-team performance spectrum.

Three contemporary examples of distributed team cognition have been articulated to inform the reader of the contexts that are complex and dynamic, and that produce risks in teamwork. Within these contexts, many possibilities and constraints make solution paths to problems either feasible or impossible to consider. Being able to conduct research using team simulations that represent scaled worlds is very important for contributing to the state-of-the-art in this area. Likewise, expanding methodological approaches to comprehensively address and triangulate these issues will produce prolific dividends. The chapter now looks at a review of research produced in this area and the trends that contribute to further comprehension.

AN OVERVIEW OF TEAM COGNITION RESEARCH—CONTEMPORARY PERSPECTIVES

When one considers the specific team simulation history coupled to work at Wright-Patterson Air Force Base as part of the entire learning constructivism, it has woven together many perspectives, engaged different theories and approaches, and compiled various measurements associated with team cognition/distributed cognition. In fact, one might say it has produced numerous published studies that provide evidence of what constitutes team cognition, what it does, how it comes about, and just as importantly how it fails. Team cognition research hence has been developed and evolved from these distant passages into the form it occupies in today's world. This history represents only one distinct constructivism in the development and formulation of distributed team cognition, albeit it is related to the extensive work of other top researchers and their associates (Nancy Cooke at Arizona State University, Mica R. Endsley at SA Technologies, Inc., Steve Fiore at the University of Central Florida, Susan Mohammed at The Pennsylvania State University, Joan Rentsch at The University of Tennessee, Ed Salas at Rice University) who have been involved in military teamwork, distributed team cognition, and collaborative technologies for the last 25 years.

Each team simulation embarked upon in the 1980s and into the late 1990s had (1) its own theoretical foundation that drove research hypotheses, as well as (2) its own ecology wherein (3) issues and problems from work settings drove questions about how cognition emerges in teams to harness efficiency and effectiveness. The niche occupied through our team simulation repertoire was definitely determined by the specific needs and requirements underlying the necessity to learn more about distributed team cognition, boundary constraints placed on these requirements, how definitions of team cognition were operationalized within the boundary constraints, and the state of technological innovation possible at the time.

The research foundation enabled a certain "instantiation" of teamwork in terms of hypotheses generated, variables of interest explored, and trajectories that have continued to influence where research headed. Inherently, the biases and focus created and influenced our own understanding of distributed team cognition in many different ways. For example, the work with TRAP was derived from the theoretical perspective of distributed dynamic decision making (Pattipati, Kleinman, & Ephrath, 1983; Wohl, 1981), whereas the work with CITIES emphasized an expansiveness of individual theories of situation awareness (Endsley, 1995), and lent credence to psychological distancing theory and the initial testing of electronically mediated communication (Santoro, 1995). Wellens' work (1993) incorporated his own theory of team situation awareness as coupled to certain types of interaction modalities and expert systems. The early work on Jasper/Repsaj tested theories related to collective induction and perceptual differentiation that afforded analogical transfer of knowledge in complex problem solving (Bransford, Brown, & Cocking, 2000; McNeese et al., 2002). These strands have come and gone but current research involving the follow-on simulation (NeoCITIES) continues to look at many elements of distributed team cognition and the impact of various forms of information and collaborative technologies.

Given this background and progression, it is now useful to look at how team research has evolved today and to consider contemporary trends, prevalent needs, and new roles for technological coupling. Stated another way, in what follows we provide an overview of what we think we know, given what we have done, and how the future could unfold.

One of the first aspects of team cognition to discuss is how it is framed within certain constraints. In the previous chapter of this volume it was established that the research at Wright-Patterson Air Force Base (that the first author was a part of) focused on the context of either C^3 or emergency crisis management, especially within the cognitive elements of decision making, problem solving, planning, and design. Although much of the work utilized experimental quantitative paradigms, it was also the case that qualitative study was used to derive understanding of complex teamwork (McNeese & Ayoub, 2011). As work expanded during the last 20 years there have been other fields of practice where distributed team cognition has become prevalent (e.g., civilian search and rescue, driving performance, human-robotic work in crisis management, medical-health informatics, natural gas processing plants, air traffic control, logistics and supply chains, cyber operations, and UAVs). Furthermore, with the development of many new distributed technologies the nature of teamwork has changed dramatically wherein technology is facilitating autonomous tasks with human and synthetic actors (McNeese et al., 2018), allowing unique distillation of patterns for awareness and activity-based responses (crowd sourcing, machine learning, fuzzy cognitive maps, citizen science), and creating seamless opportunities for information sharing and communication (e.g., Skype, Google Hangouts).

The necessity of complex problems creates demands that one person would have trouble addressing alone. Therein, the development of teams has evolved with stated rationality (i.e., intentionality/goals) wherein a member is assigned a given task where tasks (and also team members) are interdependent in their joint actions, and where people, tasks, and activities are interdependent on and distributed across the environment where work occurs. When high interdependency exists in teamwork then there are many temporal and conceptual contingencies that must be assessed and attended to. Because many contemporary problems involve cognition applied within a given context, teams frequently employ cognitive activities to accomplish their tasks. The old idiom "two heads are better than one" is certainly relevant but one must also be aware of the other idiom "too many cooks spoil the broth."

In more recent times, there have been movements (especially within military contexts) to do "more for less." For example, the use of robots or personal agents have been used to replace humans on a given team to improve productivity, practicality, or even accuracy of responses or to even do what humans cannot do (Kruijff et al., 2012). Although this strategy may prove to be successful or end up with failure, teamwork is still prevalent and needed to respond to complexity. Also, it may be the case that workers are trained to attend to multiple roles, or interchange roles given the response needed at a particular time, providing adaptive response to complexity (Marks, Sabella, Burke, & Zaccaro, 2002).

The team cognition research presented in this chapter involves team size with two to six team members, hence it is classified as small group research, rather than large group research. The presence of internet technology has made larger group research more prominent over the last ten years as different apps enable groups of people to compare their unique information to other groups of people. Therefore, people in larger groups are enabled to interact in creative ways.

Overall, the McGrath (1984) circumplex (as mentioned in the previous chapter) provides the kinds of processes relevant for distributed team cognition. Using this framework, research groups have focused on and provided in-depth studies associated with the various processes represented in the circumplex. Our research groups have emphasized processes of cognition that intersect situation awareness, mental model formation, and learning.

Challenges and Issues for Simulation

A review of team cognition literature from 1990 until the present revealed nine simulations that were utilized in at least three published articles that investigate team cognition phenomenon. As opposed to some of the early simulations, all of the more recent simulations for team cognition reviewed here were highly situated within a synthetic task environment (though some higher in fidelity than others) but none were as abstracted as the TRAP simulation discussed in the previous chapter. The simulations represent a wide variety of degrees of fidelity and ecological validity, fields of practice, ecological affordances, and methodological approaches to measuring team cognition. This section begins with a brief description of each simulation and a list of representative published studies that utilized each simulation.

SIMULATIONS IN TEAM COGNITION RESEARCH

Examples

Unsurprisingly, given the affiliation of many of team cognition researchers with military research institutions (e.g., Air Force Research Lab, Department of Defense, Office of Naval Research), many of these simulations revolve around flight, defense, and military surveillance. These simulations typically utilized undergraduate university students, although in a few cases Air Force ROTC students with some military background were recruited for participation in the distributed dynamic decision-making (DDD) and UAV simulations. The nuclear reactor simulation utilized intact control room teams. Below are brief descriptions of each simulation.

The DDD simulation uses teams of four to defend a base against incoming aircraft attacks, and was developed for the Department of Defense to use in training (Miller, Young, Kleinman, & Serfaty, 1998). DDD utilizes a grid divided into four quadrants with a base in the center. Participants use radar to detect and identify incoming aircraft as friend or foe. Each participant is assigned to a quadrant, which only he/she can monitor, and has the ability to operate four types of vehicles with different capacities for engaging incoming enemy aircraft. Because no team member can monitor, identify, or attack aircraft in other quadrants, the team members must work together to share information and coordinate attacks to protect the base. DDD has been used for measuring team cognition in many studies including Christian, Pearsall, Christian, and Ellis (2014), Ellis (2006), Ellis and Pearsall (2011), Johnson et al. (2006), Moon et al. (2004), Pearsall, Ellis, and Bell (2010), and Porter, Gogus, and Yu (2010) and to measure other team constructs outside of cognition research.

The Falcon F-16 simulation and Longbow/Gunship helicopter (AH-64 Apache) simulations utilize dyads to identify and shoot down enemy aircraft while keeping the plane/helicopter in the air. The Falcon simulation assigns a joystick to one member, who flies the plane and positions the plane for making attacks, and a keyboard to the other member, who can set airspeed, call up weapons systems, and gather additional information. Compared to flight simulators used to train pilots, this was a low fidelity simulation (Mathieu, Heffner, Goodwin, Salas, & Cannon-Bowers, 2000), but was complex compared to other commercially available PC-based simulators of the time. Team members were required to coordinate, as neither could accomplish the mission to shoot down enemy aircraft alone. Among the published papers to use the Falcon simulation to study team cognition are Chen et al. (2002), Mathieu et al. (2000), Mathieu, Heffner, Goodwin, Cannon-Bowers, and Salas (2005), and Volpe, Cannon-Bowers, and Salas (1996).

Longbow and Gunship are both low-fidelity PC-based simulators in which each dyad member utilized a joystick to operate helicopter functions; generally, one member flew the plane and fired weapons while the other operated the weapons systems, used radar for surveillance, and monitored systems. Members could not complete the task on their own as they relied on information from the other and the distributed operation of the helicopter functions to meet the goals of identifying and attacking primary and secondary targets, protecting friendly targets, and avoiding anti-aircraft attacks. Chen, Thomas, and Wallace (2005) and Marks et al. (2002) utilized the Longbow simulation, and Stout, Cannon-Bowers, Salas, and Milanovich (1999) and Stout, Salas, and Carson (1994) used the Gunship simulation.

The Uninhabited Aerial Vehicle–Synthetic Task Environment (UAV-STE or UAV, Cooke & Shope, 2004) engages teams of three members in surveillance and taking reconnaissance photos of key targets and is based on Air Force Predator operations. Team members communicate through a push-to-talk intercom system as they each execute one of three roles: pilot (who flew the vehicle, controlling airspeed, heading, and altitude), navigator (who determined flight paths and oversaw the missions), and photographer (who adjusted the camera, took photos, and monitored camera equipment). As in the previous simulations, no one member can complete a mission alone, as the interdependent roles relied on unique information from all three members in order to be successful. Examples of published articles using the UAV simulation to study team cognition include Cooke, Gorman, Duran, and Taylor (2007), Cooke, Gorman, Myers, and Duran (2013), Cooke, Kiekel, and Helm (2001), Gorman and Cooke (2011), and Gorman, Cooke, and Amazeen (2010).

Finally, in terms of military-based simulations, the Non-Combatant Evacuation Operation: Red Cross Rescue Mission (NEO) is a low fidelity, open-ended team decision-making task with three roles (weapons, intelligence, and environmental) where teams must create a rescue plan using military resources. Similar to the previous simulations, each team member received shared and unique information about the resources available for the impending rescue as well as mission goals (for example, rescue crews should avoid detection). On the other hand, teams were not constrained to specific actions as in the other simulations. Teams interacted either face-to-face or via text-based computer mediated communication to share information and develop a rescue plan. Because key information about assets was distributed across the roles, team members were interdependent on one another and communication was required. Three published articles that used this simulation to measure team mental models/schemas include McComb, Kennedy, Perryman, Warner, and Letsky (2010), Rentsch, Delise, Mello, and Staniewicz (2014), and Rentsch, Delise, Salas, and Letsky (2010).

Beyond the military context, there is additional variety of representativeness in other fields of practice. A high-fidelity gas-cooled nuclear-reactor training facility simulation on site at the nuclear plant has also been used to measure team cognition among intact plant crews. This research was not conducted in a typical research laboratory with undergraduate students, but in a training simulator used to train nuclear power plant crews in an exact replica of their working environment. As a result, the participants in these studies were intact nuclear power plant operation crews. The simulator could be programmed for a variety of situations where crews dealt with alarms, phone calls, and crisis situations. Crews members were typically separated at their stations but had the ability to talk together face-to-face when necessary. Published studies utilizing this simulation include Patrick, James, Ahmed, and Halliday (2006), Stachowski, Kaplan, and Waller (2009), and Waller, Gupta, and Giambatista (2004).

The NeoCITIES simulation (an updated version of the CITIES simulation discussed in depth in the previous chapter) is a complex crisis management simulation with three roles (EMS/fire, police, and hazmat) where teams must allocate resources in a given time period to respond to a variety of city crisis scenarios. Because CITIES was ahead of its time, NeoCITIES is still widely used. It will also be discussed in depth later in this chapter. A few examples of research utilizing the NeoCITIES simulation include Hamilton et al. (2010), Hamilton, Mancuso, Mohammed, Tesler, and McNeese (2017), Mohammed, Hamilton, Tesler, Mancuso, and McNeese (2015), Pfaff (2012), Pfaff and McNeese (2010), and Tesler, Mohammed, Hamilton, Mancuso, and McNeese (2018).

Finally, a simulation based on the game SimCity4 has also been utilized as a synthetic task environment for decision making regarding city planning and strategic management. Teams of three to four members are assigned roles to perform the management of a partially developed city created on the SimCity platform. Roles include city planning officer, financial officer, public works officer, and social welfare officer, where each role has unique information and training about the decisions that can be made within their role's purview. Therefore, team members are highly dependent on information from one another in order to meet the goal of growing the city's population. Members are given time to prioritize and plan actions, which are then put into play in the simulation. Members can see in real time how decisions affect the city's population. Some published studies utilizing the SimCity4 simulation include Randall, Resick, and DeChurch (2011), Resick, Dickson, Mitchelson, Allison, and Clark (2010), Resick, Murase, et al. (2010), and Resick, Murase, Randall, and DeChurch (2014).

Ecological Validity of the Simulations

Ecological validity varies across these simulations, although all of the simulations are situated within fields of practice with high-stakes situations where performance is vital to preserving human lives and property. The military-based simulations require participants to identify and protect against enemy aircraft, to identify and photograph enemy targets, or to plan a rescue mission. The NeoCITIES simulation puts teams in crisis management situations that range from fire and crashes to potential terrorism scenarios. The nuclear reactor control room simulation is obviously focused on maintaining the integrity of containment to protect neighboring communities. Even the SimCity4 simulation includes roles focused on the well-being of the city's population. Although the simulations are all situated in realistic, high-stakes situations, they vary in the degree to which the simulation represents the real-world context. The nuclear simulation has the most ecological validity because it is an exact replica of the control room at the plant. Simulations where participants are monitoring and flying vehicles such as DDD, Falcon, the Apache simulations, and UAV have some level of ecological validity in that they approximate the same functions that would be required by real-world teams but at a lower level of realism or complexity. In part, this is necessary if research is to be conducted using undergraduate student participants. The SimCity4 simulation allows participants to see a bird's eye view of the city they are developing, but does not immerse them in the realities of making complex decisions for a community.

However, although the realism of simulations may be questioned, the more important issue for ecological validity is whether the simulations approximate the teamwork conditions required in the real-world scenarios. Although these simulations may not have the realism of more sophisticated simulations, they have been purpose-built (or chosen, in the case of off-the-shelf software simulations like Gunship and SimCity4) to create the same kind of interdependence and requirements for communication, coordination, development of team mental models, and situational awareness that would be required in real-world teams. Mathieu et al. (2005) acknowledged the limitations of these kinds of simulations, asserting,

> Similar to the position advocated by Mathieu et al. (2000) and Marks, Zaccaro, and Mathieu (2000; Marks et al., 2002), we do not suggest that the results of this investigation would generalize directly to real-world settings such as air combat situations. Rather, we were interested in testing how teammates' mental models and team processes and performance relate to one another in general and adopted a test-bed that would enable us to do so while controlling contextual influences.

Indeed, this element of control is still a major benefit of choosing an experimental use for simulations.

Importance of Technology for Simulations

The previous chapter noted that the early days of simulation-based cognition research had an experimental focus with a desire for controlled situations and the ability to measure multiple processes and outcomes. The team cognition simulation work since 1990 retains much of that same focus. The more sophisticated simulations rely heavily on technology, specifically software that is designed to be adaptable to the needs of the research and to allow for different experimental and contextual manipulations. Many of the predominant complex simulations have been designed for flexibility, with the ability, for example, to program multiple scenarios, adjust the starting/ending points of aircraft trajectories, determine how many aircraft appear in a scenario and whether the crafts are friend or foe, or create a variety of crisis simulations and events that need responses. Because simulation specifics can be changed and controlled, simulations allow researchers to test the effects of different workloads, member rotations, unexpected events or contextual changes, use of face-to-face versus computer-mediated communication, or team member familiarity with one another (just to name a few variables). Researchers can then measure team performance, adaptability, situational awareness, communication, information sharing, transactive memory, mental models/schemas, coordination, and other team processes. Some of the simulations have been developed with the purpose of initially investigating particular types of team cognition but have also been adapted to investigate others (for example, research with the DDD simulations often measures transactive memory, but DDD has also been used in the context of measuring team interaction mental models).

Research prior to the 1990s sometimes used technology as a way to test group technologies that were being developed. That is still the case with a simulation like the NEO where the simulation itself is administered in a low-tech manner (task information provided on paper), but several software programs have been used to test the effects of that affordance on team information sharing, knowledge organization and retention, and decision making. As the nature of the workplace continues to change such that team members may be working in different locations, team cognition research continues to be interested in ways to improve sharing of information and to develop shared understanding through technology.

In many of the simulations, technology is used predominantly to deliver the content and environment of the simulation (e.g., cockpit, defense grid, vehicle controls, reactor workstations, information on incident/crisis locations). Additionally, technology is being used to enhance the richness of communication in some of the simulations, by allowing non-collocated team members to talk with one another via audio rather than just sending information to computer workstations. Additionally, technology is leveraged by many researchers to support training on task and teamwork skills and (similar to earlier research) to facilitate the collection of process and behavioral data through audio/visual recordings that are later utilized in qualitative coding.

Ecological Affordances in Simulation

Many of the simulations were designed or chosen with particular ecological affordances in the environment that can support the development of team cognition.

The major affordance in all of the simulations mentioned is the interdependency of the team members in terms of distributed information and varied abilities to perform specific behaviors. Because the tasks are designed in such a way that one team member is unable to know and do everything required for success, the simulations provide fertile ground for developing team cognition. For example, in the UAV simulation, the navigator is the only one who can determine a route to the target, but the pilot must fly it, and the photographer alone can take the required photos. To get the best photos, they must all work together within constraints and still meet the goals in an efficient manner (i.e., responding quickly, not wasting film).

Each simulation provides mechanisms for team members to gather and share their own unique information, either through the simulation software or through audio contact. Audio contact also affords teams the opportunity to elaborate on their information or discuss their strategies, which can enhance situational awareness and team mental models and, ultimately, team decision-making abilities. Affordances also include the ability to visualize information. For example, in the DDD, once the team member with an aircraft in his/her detection grid is able to identify the track, the identification information becomes visible to everyone. And, in the NEO task, teams using the CMAP or eWall visual collaboration software for communication can post and visually organize information to help the team make sense of relationships among task concepts.

Methodologies for Measuring Team and Individual Constructs

Much of the simulation research is ultimately concerned with how team cognition and processes affect team performance, hence most of the simulations discussed above include built-in objective measures of team (and often individual) performance. These quantitative measures include number of targets photographed, enemy aircraft shot down, flights successfully routed, time taken to respond to an event, number of weapons or amount of film used, money spent, or even population growth in a simulated city. Objective measures are often built into the architecture of the simulation software and are tallied without additional effort by the researchers. These are quantitative, objective measures of performance outcomes. They are relatively easy to use, can be collected the same way for every team, and provide an easy method to compare teams across conditions.

However, they do not represent the underlying team cognition of interest. In order to develop a deeper understanding of the "black box" processes of interdependent teams engaged in complex tasks, researchers must utilize qualitative methods of data collection and analysis. The most popular of methods across all of the simulations involves coding behaviors (often communications) evident in recorded audio and video of team simulation performances. Behaviors are qualitatively coded by trained experts based on pre-defined coding schemes relevant to the particular type of team cognition being studied. Coding across the duration of an exercise can allow researchers to explore temporal dimensions of team cognition processes and to determine if behavior changes with experience (across scenarios, trials, or sessions). However, although this method provides a more granular look at team phenomena it, unfortunately, captures coder perception of behaviors, but not the actual cognitions of the team members.

Hence, researchers utilizing these simulations also employ several methods to tap into team member cognitions. One popular method involves eliciting team or task mental models/schemas using a paired comparison approach; participants rate the relatedness of pairs of items (usually key dimensions or characteristics of the task or the teamwork requirements of the simulation) to create a matrix, which is then analyzed for accuracy with a subject matter expert model or for similarity with the models of other team members. However, this method is time-intensive and somewhat tedious (a paired comparison approach with ten items requires participants to make forty-five distinct ratings) and studies typically only utilize this method after the team completes the simulation, as requiring participants to complete multiple matrices during the simulation would detract from fidelity and likely exhaust subjects. Situational awareness on the other hand is often checked at multiple points during a simulation, as it can be elicited with a few questions asking participants about their understanding of the current state of affairs in a given scenario. With growing interest in communication and mental models, simulation research since the 1990s has combined objective performance scores with self-report measures of attitudes and beliefs, qualitative coding of behaviors, and participant-provided situational awareness and mental model similarity/accuracy scores.

SUSTAINED AREAS OF RESEARCH WITHIN TEAM SIMULATIONS

Individual and Team Situation Awareness

Situational awareness (SA) is long considered a fundamental teamwork construct, specifically influencing the development of both similar and accurate team cognition. The awareness that one has regarding their own team and other team member roles is often linked to team effectiveness and performance. Without adequate amounts of situational awareness at both the individual and team level, breakdowns in communication and coordination can happen which then negatively impact team cognition.

Multiple teamwork simulations have sought to specifically investigate and measure both individual and team situational awareness in the context of dynamic tasks. Mancuso, Hamilton, Tesler, Mohammed, and McNeese (2013) utilized the NeoCITIES task simulator to study situational awareness for intrusion detection analysis of cybersecurity teams. More specifically, the research team developed a new system built upon NeoCITIES, termed idsNETS. IdsNETS allowed for a task that mimicked the task of an intrusion detection analyst. Utilizing this simulation, the researchers were able to measure cyber SA, or SA that is occurring during a cyber event. Cyber SA was measured using a built-in version of the Situation Awareness Global Assessment Technique (SAGAT), which is a prominent survey metric to measure SA. In addition to SAGAT, the simulation also has a built in human-performance scoring model that calculates scores relevant to a player's response to an event in relation to overall damage. This score essentially provides a metric into understanding if players are appropriately responding to events. If they score high, then it can be assumed that they had higher levels of situational awareness. In fact, the score can be looked at in light of SAGAT to provide a more complete understanding of a player's SA. In addition to idsNETS, the regular NeoCITIES platform provides capabilities to measure individual and team SA. Using SAGAT, the simulation provides capabilities to elicit individual and team measures of both perceived and behavioral SA. In most of the NeoCITIES references found in this chapter, the measurement of SA was employed during the experiment.

Work by Gorman and colleagues (2005; Gorman, Cooke, & Winner, 2006) has also explored SA through the utilization of a simulation. Their research perspective focuses on how communication and coordination impacts the development of team-level SA. Experiments from this research team using an unmanned air vehicle task environment (UAV-STE) have sought to understand how SA is tied to coordination. Findings show that it is not necessary for every team member to be identically aware of the same information and that there is a significant need for team members to have appropriate levels of coordination in place to successfully develop or enact team SA. More recent work in this same UAV-STE has explored the role of team SA in human-autonomy teaming. The UAV-STE was originally designed for the utilization of three human team members (pilot, navigator, payload operator) to control a UAV to take photographs at specific waypoints. More recent iterations of this simulator platform have replaced the human role of the pilot with a synthetic agent. Recent work (Demir, McNeese, & Cooke, 2017) has found that pushing information (as opposed to pulling) amongst the team was positively associated with team SA. In addition, this study found that human-autonomy teams exhibited less pushing and pulling information than human-human teams. In general, much like human-human teams, this study highlights that the anticipation of team member behaviors is important in team SA. In similar work, McNeese and colleagues (2018) found that human-human and human-autonomy teams had statically equivalent levels of team situational awareness. This is an important finding as it shows potential from human-autonomy teams to operate in an effective manner. Traditionally speaking, some researchers think that a human-autonomy team could not exhibit effective levels of performance on multiple teamwork dimensions. This may have some truth to it, but this work shows that it is possible for a human-autonomy team to, at a minimum, develop teamwork manners equal to some human-human teams.

Team Schemas

Team schemas have been studied using more complex simulations that tend to require team members to take on expert roles and to acquire substantial information prior to being able to execute the team task. Mancuso et al. (2013) employed an emergency crisis management simulation known as the NeoCITIES 3.1 simulation. The simulation involved scenarios generated from observations of and interviews with individuals on actual crisis management teams. Three-member teams with each member assigned to one of three specific roles where each role has specific resources and functions were to respond to two scenarios containing eighteen events. Successful performance required team members to assess the problem and to communicate and coordinate with each other. Team members worked together via technology designed to simulate a dispatch terminal. This complex simulation required training lasting approximately five minutes followed by a five-minute training scenario as practice. The NeoCITIES scenarios were approximately fifteen minutes each. Teams completed two scenarios which could be scripted to varying levels of difficulty. High performance scores were obtained when team members employed appropriate resources. Scenarios were designed such that teams had to be judicious in allocating resources. Thus, the task was scorable.

Mohammed and Ringseis (2001) used a modified version of the Tower Market Task (Beggs, Brett, & Weingart, 1989) which included four stores (grocery, florist, bakery, and liquor store) and each group member served as a representative of one store. The representatives worked together to make business decisions of mutual interest for the stores (e.g., advertising, building temperature). There was no right answer for this version of the simulation. Participants received common information and unique information regarding the store they were representing that included preferences for the decisions to be made by the group. Once groups of representatives were formed to make four mutual decisions, they had thirty minutes to achieve answers.

Rentsch et al. (2014) and Rentsch et al. (2010) utilized a simulation developed by the United States Navy based on a noncombatant evacuation operation (Biron, Burkman, & Warner, 2008). In the simulation, each of three team members was assigned an expert role. Team members were provided common information regarding the goals of the mission and unique information associated with their expert roles. In order to successfully plan the rescue, team members had to share their unique information. Task plans were scorable. Each role contained substantial information such that team members were given time to review their information packets and then teams were allotted sixty minutes to develop their rescue plan.

The complexity of these simulations presents a double edge. The benefits include increased realism and holding participants' interest. The complexity, in some cases, requires teams to spend more time on the task which enables team processes to unfold and reveal themselves. The disadvantage of the complexity is the time required for participants to adequately acquire the knowledge needed to complete the task. We contend there is a sweet spot for task complexity that enables tests for fully developed team processes without introducing error associated with team members' lack of understanding of the task. These tasks can also be designed to be scorable and afford the analysis of process data and qualitative observations.

Information Sharing

Samples of simulations employed in the study of information sharing among team members include decision-making dilemmas and hidden profile tasks. Dubrovsky, Kiesler, and Sethna (1991) used a dilemma decision-making simulation consisting of four real-life decisions around issues relevant to college students (e.g., career choices and freshman curriculum). Each decision offered an attractive but risky option versus a less attractive but safer option. Groups were given about fifteen minutes to reach consensus for each decision. There were no correct answers.

In his studies of the information-sampling model, Stasser and his colleagues favored hidden profile simulations in which all group members possessed common information and each member possessed unique information critical to identifying the problem solution. In order to achieve high-quality solutions, groups working with hidden profile tasks discover the correct problem solution only when all members share and discuss all relevant information (i.e., members must pool their information). Stasser and Stewart (1992) constructed a hidden profile using a murder mystery in which task information was presented in a twenty-seven-page booklet that included interviews conducted as part of a murder investigation. Twenty-four clues identifying which of three murder suspects was guilty were embedded in the interviews. The interviews (and therefore the clues) were distributed among group members such that groups would have access to all clues only if each member shared all his or her available clues. Combining the clues presented a clear case identifying one suspect as guilty. Group members were provided time to review the task materials and then they met together to discuss the problem and determine the primary suspect. Stasser, Vaughan, and Stewart (2000) used a similar type of hidden profile task in which group members were to complete a decision-making simulation by evaluating characteristics of three candidates running for student-body president.

These simulations have several commendable characteristics. First, they are straightforward simulations requiring no specialized knowledge, making them appropriate for student participants and relatively short data collection periods. These features also make them amenable to holding participants' interest (e.g., Dubrovsky et al., 1991). Second, these simulations tend to contain standardized, static information (e.g., Stasser et al., 2000). Third, they lend themselves to quantitative and qualitative data collections. Fourth, these types of simulations can be designed to offer closed-ended solutions that are scorable or offer preference options that are not scorable. Fifth, such simulations may be designed to include manipulations such as distributing information to create "expert" team members (e.g., Stasser et al., 2000).

One limitation of these simulations is that they tend to be quite static, which is appropriate for testing many hypotheses, but may limit their usefulness in testing hypotheses where dynamic simulations may be more appropriate. It is very important to select a simulation containing the characteristics (e.g., an available correct solution versus no correct solution available; static versus dynamic) that maximize researchers' ability to adequately test their hypotheses.

We now take a look at some of the trends within distributed team cognition and what they mean.

Trends—Strengths and Limitations

Teamwork simulations have contributed to many of the current insights we have obtained regarding team cognition, information sharing, team situational awareness, and a myriad of other team variables. In reviewing the teamwork simulation research for this chapter, we have identified some general trends that point towards both strengths and limitations of teamwork literature.

One positive trend is the utilization of teamwork simulators that are developed with insight from relevant contexts. In this chapter, we have identified multiple relevant contexts that are linked to simulators, such as command and control, emergency management crisis, and cybersecurity. These are all contexts that are (1) deeply important to societal impact and (2) highly dependent on teamwork to adequately function. In addition, another positive trend that is found in most teamwork simulators is the collection of various types of teamwork-related data. For example, most of these simulators collect data relevant to both the individual and the team on dimensions such as performance, communication, coordination, SA, and workload. Taken together these data provide a more holistic sense of what a team is doing and how effective they are.

Some of the limitations that we identified relate to a lack of qualitative work being used in concert with these simulators. In general, it is necessary for more qualitative work to be conducted in the teamwork literature. It is one thing to say that team X is performing at this level or has this amount of SA, but more is needed beyond that. The research community needs to continue collecting this type of data, but we need to also do a better job of asking participants why and how they think they performed in such a manner. And, we are not just referring to survey work as being adequate to fill the lack of qualitative data. We need interviews, focus groups, concept mapping, etc., to occur in addition to our traditional data collection methods.

Another limitation to most of the current work is a lack of simulators that accurately simulate real consequences of the work environment. For example, a UAV simulator is not simulating real-world pressure and the consequences of real-world failure. Simulations need to better represent things like time pressure, stress, and fatigue. There is some work that attempts to adequately simulate this but much more needs to be done to attempt to make the simulation environment feel more realistic beyond just the task being contextually relevant. Finally, we feel that the work conducted in the lab using a simulator needs to be better utilized and introduced back into the real environment that the simulator is indeed simulating. The research results that are found in the lab must be then implemented or at least considered in a real-world environment with the aim to improve teamwork in that environment. We recommend using the living lab approach to better implement research back into a real context (McNeese, Perusich, & Rentsch, 2000).

A CONCEPTUAL FRAMEWORK FOR DISTRIBUTED TEAM COGNITION AND SIMULATION

Given all the research conducted on distributed team cognition using differing theories, variables of interest, methods, measurements, contexts, and technologies it is useful to consider a common framework to establish (1) an integrative knowledge of the phenomena under study, (2) a plan to execute specific research requirements and approaches to enable contributions that help formulate integrative knowledge, and (3) a means for specifying and developing what a team simulation should consist of, in order to address the underlying requirements. From the mid-1990s forward to 2016 the senior author's research group created a framework that satisfies these objectives—termed the Living Laboratory Framework (LLF; McNeese, 1996, 2004; McNeese et al., 2005; McNeese & Forster, 2017).

The framework emerged from transdisciplinary research (much of which was described in the previous chapter) with the goal of merging research with design, humans with technology, and making context a basis for understanding cognition when complex systems are part of a field of practice. The basis for this framework derived from the first author's work in making sense of socio-cognitive systems from ecological viewpoints (see McNeese, 1992; Young & McNeese, 1995) where problem solving and decision making are prominent in social learning and cooperative teamwork. The foundation in this area came from the IDEAL problem solver model (Bransford & Stein, 1984) which is an early proclamation of problem-based learning.

The heart of the Living Laboratory Framework proposes problem-based learning (PBL; Norman & Schmidt, 1992) as a core value in determining research intentionality. PBL as a social learning experience encompasses constructivism where practice, reflection, and insights are primary constituents of comprehending the phenomena under study. Hence, the framework would be considered a problem-centric approach to transformative research within distributed team cognition. The framework essentially considers the trichotomy of theory à problems ß practice as informed by four interrelated research acts. It is self-directed, employs opportunistic problem finding and exploration, emphasizes collaborative knowledge seeking, and is situated for a given context of interest (e.g., a field of practice such as emergency crisis management) (Dochy, Segers, Van den Bossche, & Gijbels, 2003; Hmelo-Silver, 2004; Schmidt & Moust, 2000). The LLF (see McNeese & Forster, 2017) hence is the embodiment of PBL with the intent to broadly address many facets of understanding distributed team cognition through an integration of multiple methodological approaches. The use of the LLF enables a researcher to answer many of the conceptual questions posed in the introductory section that enable meaningful team simulations for the purpose at hand.

The original framework (McNeese, 1996) had a central node emanating directly from the PBL foundations—identify and define problems—whereupon problems could jointly be addressed through theory and practice. The LLF considers theory and practice as underlying problem-based learning. In the original framework problem-based learning is bolstered (surrounded) by four inter-connected activities to explore and develop increased understanding of distributed team cognition: (1) knowledge elicitation, (2) ethnographic study, (3) scaled world simulation, and (4) reconfigurable prototypes. (See Figure 3.1 below.) The model assumes that the input and output of these activities are flexible but that together they buildup the knowledge base to address the problems under consideration (i.e., for a phenomenon under consideration in distributed team cognition).

A basic tenet in these activities is that research is a form of continuous process improvement wherein feedback and feedforward processes are critical for both iteration and extensibility. The LLF can be activated iteratively (and thereby achieve an acceptable level of depth for each activity) to produce the best possible scaled world simulation within the constraints that are active, that addresses the specific research problem at hand.

Although an optimal situation would afford enough time and effort to thoroughly analyze each of these activities in-depth, the reality of practical situations is that only a subset of resources may be available for use to the researcher. In these cases, a researcher must make sound decisions and establish priorities for the kind of information that is going to help facilitate the most beneficial problem finding/solving combination.

These research activities can be informed from a top-down (theory-centered) or bottom-up (practice-centered) approach to begin initiation of the research cycle. A bottom-up approach would generally begin with ethnography or knowledge elicitation activity, whereas a top-down approach would begin with simulation using scaled worlds for establishing experimental-quantitative studies. The activities also highlight a "middle-through" perspective with the output being the production of a design that addresses the issues resident in the problem that was defined as the anchor of the process to begin with. Prototypes are the seeds that are translated into actual designs that would then be tested in the field along with requisite feedback for improvement. The idea of the LLF is to provide dynamic feedback from these activities to supply a holistic understanding of complex phenomena and in turn address issues, tradeoffs, and conflicts within defined problem states.

INTEGRATIVE LIVING LABORATORY EXTENDED FRAMEWORK FOR FUTURISTIC TECHNOLOGIES (ILEFT)

In 2017 the LLF was updated and extended (see Figure 3.1) from the basic model to a more refined research perspective that additionally emphasized the use of data science techniques to further inform theory-problem-practice. The other new component of the extended Living Lab is that user and researcher experience form an orchestrated mutuality of understanding that affords feedback and feedforward experiences. In many current approaches, only user experience is elicited, but the researcher's knowledge and experience is just as important to provide progressive learning about phenomena. The researcher is an active participant in the Living Laboratory, not just a passive component gaining knowledge from users. The new perspective provided the means to address user experience and researcher experience in dynamic worlds that were heavily coupled with advanced technological systems (such as artificial intelligence, robotics, and social media).

For this chapter, another update has been added to the LLF, primarily the incorporation of the fifth activity: testing, evaluation, and validation. It is important to provide triangulation among the various activity components to test, evaluate, and validate to the extent possible the veridicality of data. This solidifies findings and helps formulate an integrative, holistic nature of findings using multiple methods. Note that each of the activity components can use different methods or techniques to accomplish the purpose of the activity being addressed (e.g., knowledge elicitation may be accomplished through interviews or concept maps). For the extended framework there is also a continued emphasis on opportunities to employ broad datasets (what is now often referred to as "data science") whenever possible. This extended

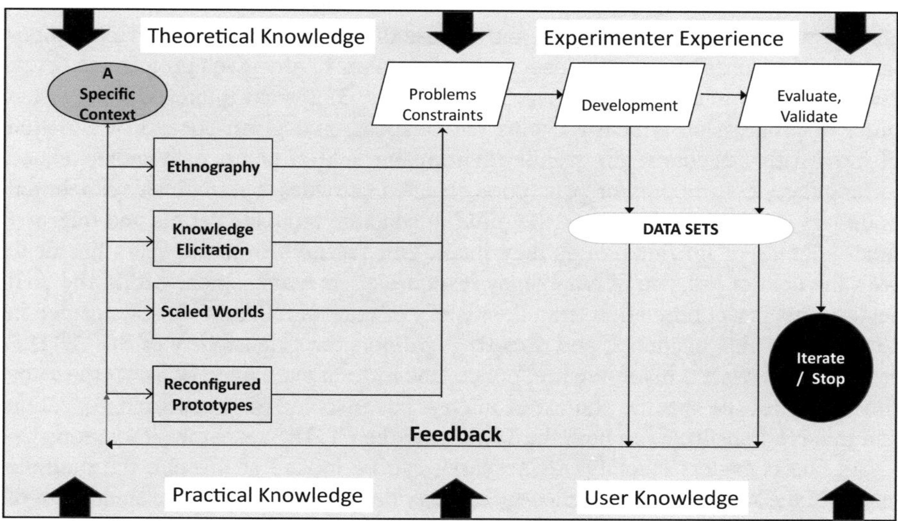

FIGURE 3.1 The Integrative Living Laboratory Extended Framework for Futuristic Technologies.

Source: Adapted from McNeese, 2017.

framework amplifies the idea that online teamwork can produce the availability of large sets of data (referred to as "big data") that can be utilized with predictive analytics to inform decision making and in particular distill meaningful patterns that are representative of issues, constraints, and tradeoffs in a given problem space.

The application of the LLF has been used many times for various applications and fields of practice to solve problems, develop technologies, and to investigate cognitive phenomena from a multi-dimensional perspective. It has been used for research involving team performance in C^3 battle management, emergency crisis management, aviation, cybersecurity, police cognition, and health and medicine (McNeese & Forster, 2017). Some of the technology design applications emanating from the LLF are geo-collaborative information systems, predictive decision aids, SA decision aids, fuzzy cognitive maps, intelligent group interfaces, team-based agent architectures, extensible team simulation architectures, and an electronically mediated communication suite. To help explain the use of the framework, a case of application is provided.

Application of Framework to a Specific Case

NeoCITIES Research

This case is beneficial for demonstrating the value of the original LLF and it places emphasis on iteration, extensible production, and continuous process improvement, and shows how the LLF functions as an interdisciplinary, transformative framework. Figure 3.1 is used to show a given temporal progression in using LLF. One can begin about anywhere in the framework and the use of the concepts and activity components is in accordance with the needs and constraints of any given project. However, owing to the framework being heavily coupled to ecological interests, a suggested order of progression typically begins with a focus on a given context or situation of interest that occurs in an overall environment/field of practice. Whether related to language, culture, and/or perception, context provides a researcher with initial boundary constraints that allow the mind to generate requisite details and interpret other sources of information as they incur. Our research over the years has dealt with the field of practice of emergency response/crisis management within the civil sector. This is a continuation from the history of team simulation work described in Chapter 1 of this handbook and directly highlights the extensibility of the CITIES research as a context involving fire, police, and hazmat team operations. At the same time, much of the specific context of interest has involved team resource allocation and this is extensible from both the TRAP and the CITIES research simulations.

As one considers NeoCITIES, research can be looked at through the multiple lenses of theory-problems-practice-technology development. From the standpoint of research interests—and specifically the work at Penn State University over the last 15 years—we have been interested in team cognition specifically related to team mental models and team situation awareness, but also how teams end up sharing and communicating information necessary to solve problems and make decisions. That interest directly segues into technology developments as much of the research crisscrosses electronically mediated communications and artificial intelligence. In fact, this was Wellens' basic interest as his original research (Wellens, 1986, 1993) examined how team SA and communications functioned when teams were coupled through video, text, and phone interfaces. The notion of distributed cognition is highly relevant in what we study wherein information-communication-collaboration technologies can help facilitate effective sharing of information when teams are distributed across environments. In today's world where Twitter, texts, and social media run rampant, distributed communication and cognition are omnipresent and taken for granted. Therefore, the research stream considered team cognition using different technologies to see whether team mental models and situation awareness could be as good as or better than face-to-face communication. This is the basic level of our research approach. Using the LLF to investigate different facets of this question has been highly valuable and it has been iterated and elaborated on many levels. The following describes the way NeoCITIES has been developed into maturity during the last few years using the LLF elements in Figure 3.1. The approach taken begins by exploring the context starting with the bottom-up approach. Ethnography and knowledge elicitation are the primary activities employed.

Ethnography

NeoCITIES incorporates many of the basics within the old CITIES paradigm inclusive of the crisis management focus. But as time has marched on it was important to achieve more veridical information to anchor the research to realistic contexts of information in order to add specificity to events, situations, and operational reality. Ethnography can be applied in different ways but over the years our research has engaged with operations centers or been able to observe certain exercises within relevant fields of practice (e.g., police cognition, Glantz, 2017; emergency medical services, Jones & McNeese, 2006; 911 dispatch centers, Terrell, McNeese, & Jefferson, 2004). Problems have been overturned and explored and meaning has been derived through observing the environment where work takes shape. When working with operational exercises, ethnography tends to emerge as observation in the moment wherein observers take notes to document human involvement with activities in a given situation, and how they may come to use the available technologies they have. For command center settings, it may be a planned approach to layer observations with interviews or use of other methods (e.g., concept mapping or cognitive task analysis). Because organizations often do not just open their doors for extended ethnography/interviews, one may have to compact these activities into a day or two and capture what one can within constraining conditions. In fact, there is a lot of pre-ethnographic setup effort that is often needed with getting access to a company or government entity to enable ethnographic analysis. In reference to these kind of constraints versus outcome possibilities. Whitaker, Selvaraj, Brown, and McNeese (1995) point out that,

> Collecting observational data on real-world design activities is time-consuming, resource-intensive work. On the other hand, there is no other method guaranteed to provide so detailed a trace of design collaborations. The scale of investment necessary to conduct an observation study should motivate researchers to comprehensively plan their tactics and budget their resources. Our experience affirms the utility of observational approaches in studying design collaboration. Recent advocacy of such methods (especially in academic circles), is well justified but solid "how-to" information and/or experienced observational researchers are less readily obtained.
>
> **(p. 46)**

Ethnographic research is important to lay a foundation for the cognate areas being addressed and most importantly can reveal problems in context that may be hidden when applying other techniques. Yet as stated above, it must be highly planned and done in a pragmatic way. For NeoCITIES, we have had the opportunity to engage through various observations and interviews, uncovering valuable knowledge which informed problem definition, information valence, and most importantly the development of scenario richness and specificity. This helps to ensure the validity of what we are studying and helps to make simulations "scaled worlds" rather than just shallow, experimental toy worlds.

Knowledge Elicitation

Once the foundation is laid for comprehending a given context within a field of practice, a researcher can begin working with subject matter experts (and even novices) to elicit more knowledge about the domain and perhaps engage in sensemaking (Klein, Moon, & Hoffman, 2006) that helps situate observations captured in the previous elements. Knowledge elicitation consists of finding relevant experts acting in the environment/context of interest, working with them to acquire knowledge through various means, then representing knowledge for future use to aggregate further comprehension and learning. The outcomes of ethnography and knowledge elicitation are strengthened when they receive feedback from each other as this amplifies the interpretive component of research investigation. Knowledge elicitation may take various forms such as survey design, interviews, concept maps, concept grids, or schemas, (see Cooke, 1994, for an in-depth review). Dependent on what is desired as a knowledge format and how much time the elicitation is limited to, different techniques may be necessitated. For many of our projects, the use of concept mapping has been used to advantage and forms the basis of the elicitation process and representation language used to capture expertise (Zaff et al., 1993; McNeese, Zaff, Citera, Brown, & Whitaker, 1995) for different fields of practice. Elicitation of knowledge along with results of ethnography may serve as a bridge to validate prior theoretical knowledge or current results from experimental quantitative data, hence establishing a triangulation of data, methods, and theory (Tiainen & Koivunen, 2006). This is an example of triangulated interpretation that strengthens reliability and validity of what is known and helps to establish research integrity. It is also a bridge in the sense that knowledge elicited may directly fold back into the scaled world simulation either as scenario development, or for informing specifics of the demands of cognitive tasks, or the temporal-procedural aspects of carrying forth an action.

As applied to crisis management research, the senior author's group has had opportunities to interview novices and experts in this domain and related domains (e.g., intelligence analysis work, battle management operations) that provide information for clarifying the contextual demands, inform theory and experiments, contribute to scenario development and design, and leverage the basis for technological innovation and design. As mentioned, the need of every research situation is different and the LLF can be adapted to meet these needs. The representation typology used to capture knowledge obtained is primarily in the form of declarative and procedural concept maps (Moon, Hoffman, Novak, & Canus, 2011; Zaff et al., 1993) often obtained from both individuals and team members for contrast and comparison.

Scaled World Simulation

Just as bottom-up knowledge from the environment and subject matter experts are informative for contextualizing theories and research experiments and creating useful scenarios, scaled world development is critical for realistically representing processes that exist in an environment. The key is to be able to place them into a simulation testbed that is both replicable and controllable. One hopes to maintain the fidelity and richness of the environment to the extent possible (i.e., to make it a scaled world). A scaled world should be a practical experimental venue that has much of the operative knowledge, contextual variation, and situated complexity that is extant in the situation the scaled world purports to model. It should also be flexible, adaptable, and extensible in that the interface, systems architecture, technological milieu, and the sense surround jointly act to portray realistic work settings.

NeoCITIES is a scaled world simulation that has evolved and been adapted for use in many experiments (McNeese & Forster, 2017). It is essentially the same team task as CITIES described in the previous chapter but housed in a client-server architecture for maximal adaptation and control. As communicated in the previous section it has evolved somewhat differently over the years as a function of the scenarios developed with the help of knowledge elicitation and expert judgments. As related to LLF, scenarios capture the problem elements and issues that need resolved. Scenarios can be thought of as complex stories which have various degrees of ill-definition present. Oftentimes the stories connect to the theoretical phenomena to be studied in that they require (1) forms of team situation awareness, (2) shared understanding of teamwork (team schemas, team mental models), (3) ways of addressing information overload and temporal requirements, and (4) coordinating resources across members at the right point in time to yield the greatest outcome (distributed information spaces). The stories that underlie actual data presented contain the elements of situations and events that unfold across time.

The primary team task, team resource allocation, is based on each team member allocating resources (under their control) to alleviate the problems/issues inherent in the situation. For a given situation, the demand may be for one, two, or three members to respond to an event appropriately with the right type and number of resources to resolve the situation. If a situation escalates then the team begins to lose points (performance decrements). NeoCITIES typically consists of a three-person team wherein each member performs a different role: police, fire, or hazmat operations.

Each role has a specialized but limited number of resources (e.g., five police squad cars, five fire emergency response vehicles to begin the simulation) which may be allocated within a set timeframe. The simulation is emergent in that differing situations and events pop up within the time constraints of the entire simulation. Team members must exchange timely information as needed (using distributed chatrooms) in order to enhance coordination, collaboration, and communication. Research has included the derived performance score as a pertinent dependent variable while communications across the team are captured for further analyses (see, for example, Pfaff, 2012). Additionally, many independent and dependent variables have been incorporated based on the phenomena under study (Hamilton et al., 2017).

An important part of a scaled world simulation is the ability to provide interfaces to technological products that are embedded for testing within the simulation. NeoCITIES has been very flexible for this purpose and various technological innovations have been the focus of experimental investigation (e.g., fuzzy cognitive maps, Jones, 2006; intelligent group interfaces, Connors, 2006; and team-computer interfaces, Hellar & McNeese, 2010).

Reconfigurable Prototypes

One point of emphasis in this handbook is the integration of new technologies within distributed cognition contexts. This particular element provides the basis of developing with designs as prototypes which can be informed by the bottom-up processes, with the output of a prototype being the basis for assuaging the problem identified and represented in the LLF. The idea of "reconfigurable" connotes that designs are introduced as transmutable ideas, developed through an iterative fashion into working technologies (i.e., a prototype). Making a prototype reconfigurable helps to ensure there is a good fit between team processes and the tool or technology being designed. The prototype can be tested in various ways by users themselves or as an active component in the simulation to further refine its user-centeredness, hence making it reflect both researcher and user experience. The prototype is evolved through use by targeted users (naïve and expert) contributing to the ecological validity of the design as an intended solution to the problem at hand. By iterating studies over time with continuous process improvement, effective user-centered designs emerge and can be validated against the problem state. The scenarios used in the simulation can also provide specific timelines and demands to determine what attributes and information fields need to be present to support user functionality.

When applied to the NeoCITIES research case, reconfigurable prototyping has been valuable for developing the team interface (i.e., the human-computer interface for the team(s) participating in the simulation). Early designs developed for the NeoCITIES simulation employed the AKADAM techniques wherein concept mapping elements were translated into design elements using a timeline-based design storyboarding (see McNeese et al., 1995). This is evidenced by various formulations generated first as prototypes and then instantiated as an actual interface designed for individual roles and team use. Some of these interfaces emphasized more perceptual elements (e.g., geography orientation, map-based awareness, visual arrays) whereas others were based on typical human-computer interaction frames that were more verbal-language oriented, and many interfaces utilized both perspectives. Regardless, both were reliant upon chat room technology for team communication.

One open-source programming environment the senior author's group has utilized to facilitate prototype-to-use applications within client-server architectures is described by Hamilton et al. (2010):

> The newest iteration, NeoCITIES 3.0, was built using Web 2.0 technologies to be a more flexible and distributable application. It was designed using a Model-View-Controller (MVC) architecture and built using Java, Adobe BlazeDS and Adobe Flex technologies (Hellar, 2009). NeoCITIES 3.0 is a cross-platform application (e.g., Windows, Mac, Linux) that runs on an Apache Web Server within a Java Virtual machine. It has afforded much adaptability, the production of engaging interface designs, and yields fast mockups of interfaces for the simulation.
>
> **(p. 433)**

In addition to developing interfaces for the simulation per se, prototyping may be used to develop other design innovations (e.g., aids, virtual environment adaptation, information displays) that require information and communications technologies (McNeese, Brewer, Jones, & Connors, 2006). Typically, these designs become independent variables that are part of a test and evaluation protocol to determine whether they enhance distributed team cognition/communication in solving a given scenario within the simulation. Examples of these kinds of designs would be intelligent group interfaces (Connor, 2006), fuzzy cognitive map aids (Jones, 2006), and predictive attention aids (Minotra & McNeese, 2017).

Extensible Simulations

One of the major advantages of the NeoCITIES series of simulations is that the base architecture can be adapted for other potential domains that contain a given form of resource allocation and situation assessment as primary task requirements. Scenarios contain stories and problems that are indigenous to the domain under consideration and are constructed in an iterative manner. New scenarios provide new content that is relevant for a scaled world representative of the new domain. Examples of extensible adaptations are intelligence operations incorporating team-to-team transactions (Connors, 2006), cybersecurity (network access) simulation (Reiffers, 2010), distributed cyber teams (Mancuso, Minotra, Giacobe, McNeese, & Tyworth, 2012), and cybersecurity event monitoring (Minotra, 2012). Although domain extensibility involves more investment to acquire and refine legitimate knowledge for scenario development, the basic NeoCITIES focus in emergency crisis management has been iterated, transformed, and reified many times for unique experimental studies involving elements of distributed team cognition (e.g., see Pfaff & McNeese, 2010 for study of moods and stress within teamwork; Endsley, 2016, for how culture impacts team performance; Tesler et al., 2018, for the effects of storytelling and guided team reflexivity on teamwork).

CONCLUDING REMARKS

Over the years the ideas behind team cognition and distributed cognition have been reified and tested through the development of team simulations, as utilized by various researchers testing many aspects of team process and performance, leading to valuable contributions and insights. In certain cases, these insights have informed design resulting in the proliferation of useful technologies that help to assist, enable, or amplify cognition at the individual and team level. In other cases, the simulations provide a nexus that contributes to emergence of team and distributed cognition being blurred (therein the term *distributed team cognition*). As has been pointed out, the Living Laboratory Framework has been a faithful and extremely useful mechanism in taking a holistic approach to understanding the interplay among humans, technologies, and their environments. It has provided a basic-level approach in determining how to go about implementing a team simulation that has a good chance of improving understanding of the crossroads of psychological processes, ecological niches, and technological innovation. Team simulation holds great promise for future progress in understanding and responding to the extant needs within distributed team cognition. Hopefully, this chapter will make that promise come true sooner rather than later.

NOTE

1. The MINDS group originally was established at Pennsylvania State University as a laboratory and interdisciplinary research group which specifically utilized the acronym MINDS as representative of Multidisciplinary Initiatives in Naturalistic Decision-Making Systems. The group focuses on conducting interdisciplinary research studies that integrate information, people, technologies, and contexts in a meaningful nexus.

REFERENCES

Beggs, R. J., Brett, J. M., & Weingart, L. R. (1989). *Teaching note for Towers Market*. Evanston, IL: Northwestern University, Dispute Resolution Research Center.

Biron, H. C., Burkman, L. M., & Warner, N. (2008, April 15). *A re-analysis of the collaborative knowledge transcripts from a non-combatant evacuation operation scenario:*

The next phase in the evolution of a team collaboration model. Naval Air Systems Command Technical Report, NAWCADPAX/TR-2008/43. Patuxent River, MD: Naval Air Warfare Center.

Bransford, J. D., Brown, A. L., & Cocking, R. R. (Eds.). (2000). *How people learn: Brain, mind, experience and school.* Washington, DC: National Academies Press.

Bransford, J., & Stein, B. (1984). *The IDEAL problem solver.* New York: W. H. Freeman.

Brown, J. S., Collins, A., & Duguid, P. (1989). Situated cognition and the culture of learning. *Educational Researcher, 18*(1), 32–41.

Chen, G., Thomas, B., & Wallace, J. C. (2005). A multilevel examination of the relationships among training outcomes, mediating regulatory processes, and adaptive performance. *Journal of Applied Psychology, 90*(5), 827–841.

Chen, G., Webber, S. S., Bliese, P. D., Mathieu, J. E., Payne, S. C., Born, D. H., & Zaccaro, S. J. (2002). Simultaneous examination of the antecedent and consequences of efficacy beliefs at multiple levels of analysis. *Human Performance, 15*(4), 381–409.

Christian, J. S., Pearsall, M. J., Christian, M. S., & Ellis, A. P. J. (2014). Exploring the benefits and boundaries of transactive memory systems in adapting to team member loss. *Group Dynamics: Theory, Research, and Practice, 18*(1), 69–86.

Clapway (n.d.). *Grandfather of Virtual Reality Embarks on New Adventure*, available at https://clapway.com/2015/07/18/grandfather-of-virtual-reality-embarks-on-new-adventure-233/ (accessed April 22, 2020).

Connors, E. S. (2006). *Intelligent group interfaces: Envisioned designs for exploring team cognition in emergency crisis management.* Unpublished doctoral dissertation, The Pennsylvania State University, University Park, PA.

Cooke, N. J. (1994). Varieties of knowledge elicitation. *International Journal of Human-Computer Studies, 41*(6), 801–849.

Cooke, N. J., Gorman, J. C., Duran, J. L., & Taylor, A. R. (2007). Team cognition in experienced command-and-control teams. *Journal of Experimental Psychology: Applied, 13*(3), 146–157.

Cooke, N. J., Gorman, J. C., Myers, C. W., & Duran, J. L. (2013). Interactive team cognition. *Cognitive Science, 37*, 255–285.

Cooke, N. J., Kiekel, P. A., & Helm, E. (2001). Measuring team knowledge during skill acquisition of a complex task. *International Journal of Cognitive Ergonomics: Special Section on Knowledge Acquisition, 5*, 297–315.

Cooke, N. J., & Shope, S. M. (2002). The CERTT-UAV task: A synthetic task environment to facilitate team research. In *Proceedings of the Advanced Simulation Technologies Conference: Military, Government, and Aerospace Simulation symposium* (pp. 25–30). San Diego, CA: The Society for Modeling and Simulation International.

Cooke, N. J., & Shope, S. M. (2004). Designing a synthetic task environment. In S. G. Schiflett, L. R. Elliott, E. Salas, & M. D. Coovert (Eds.), *Scaled worlds: Development, validation, and application* (pp. 263–278). Surrey, UK: Ashgate.

Demir, M., McNeese, N. J., & Cooke, N. J. (2017). Team situation awareness within the context of human-autonomy teaming. *Cognitive Systems Research, 46*, 3–12.

Dochy, F., Segers, M., Van den Bossche, P., & Gijbels, D. (2003). Effects of problem-based learning: A meta-analysis. *Learning and Instruction, 13*(5), 533–568.

Dubrovsky, V., Kiesler, S., & Sethna, B. (1991). The equalization phenomenon: Status effects in computer-mediated and face-to-face decision making groups. *Human Computer Interaction, 6*, 119–146.

Ellis, A. P. J. (2006). System breakdown: The role of mental models and transactive memory in the relationship between acute stress and team performance. *Academy of Management Journal, 49*(3), 576–589.

Ellis, A. P. J., & Pearsall, M. J. (2011). Reducing the negative effects of stress in teams through cross-training: A job demands-resources model. *Group Dynamics: Theory, Research, and Practice*, *15*(1), 16–31.

Endsley, M. R. (1995). Measurement of situation awareness in dynamic systems. *Human Factors*, *37*(1), 65–84.

Endsley, T. C. (2016). *An examination of cultural influences on team cognition and information sharing in emergency crisis management domains: A mixed methodology approach.* Unpublished doctoral dissertation, The Pennsylvania State University, University Park, PA.

Fan, X., McNeese, M., Sun, B., Hanratty, T., Allender, L., & Yen, J. (2009). Human-agent collaboration for time stressed multi-context decision making. *IEEE Transactions on Systems, Man, and Cybernetics (A)*, *40*(2), 306–320.

Gao, F., Cummings, M. L., & Solovey, E. (2014). Modeling teamwork in supervisory control of multiple robots. *IEEE Transactions on Systems, Man, and Cybernetics (B)*, *44*(4), 441–453.

Gibson, J. J. (1979). *The ecological approach to visual perception.* Boston, MA: Houghton Mifflin.

Glantz, E. (2017). Police cognition and participatory design. In M. D. McNeese & P. K. Forster (Eds.), *Cognitive systems engineering: An integrative living laboratory framework* (pp. 299–312). Boca Raton, FL: CRC Taylor & Francis Publishing, Inc.

Gorman, J. C., & Cooke, N. J. (2011). Changes in team cognition after a retention interval: The benefits of mixing it up. *Journal of Experimental Psychology: Applied*, *17*, 303–319.

Gorman, J. C., Cooke, N. J., & Amazeen, P. G. (2010). Training adaptive teams. *Human Factors*, *52*, 295–307.

Gorman, J. C., Cooke, N. J., Pederson, H. K., & DeJoode, J. A. (2005, September). Coordinated Awareness of Situation by Teams (CAST): Measuring team situation awareness of a communication glitch. In *Proceedings of the Human Factors and Ergonomics Society annual meeting* (Vol. 49, No. 3, pp. 274–277). Los Angeles, CA: SAGE Publications.

Gorman, J. C., Cooke, N. J., & Winner, J. L. (2006). Measuring team situation awareness in decentralized command and control environments. *Ergonomics*, *49*(12–13), 1312–1325.

Greeno, J. G. (1998). The situativity of knowing, learning, and research. *American Psychologist*, *53*(1), 5–26.

Gross, T., Stary, C., & Totter, A. (2005). User-centered awareness in computer-supported cooperative work-systems: Structured embedding of findings from social sciences. *International Journal of Human-Computer Interaction*, *18*(3), 323–360.

Hamilton, K., Mancuso, V., Minotra, D., Hoult, R., Mohammed, S., Parr, A., . . . McNeese, M. (2010). Using the NeoCITIES 3.1 simulation to study and measure team cognition. In *Proceedings of the 54th annual meeting of the Human Factors and Ergonomics Society* (pp. 433–437). San Francisco, CA: Human Factors and Ergonomics Society.

Hamilton, K., Mancuso, V., Mohammed, S., Tesler, R., & McNeese, M. (2017). Skilled and unaware: The interactive effects of team cognition, team metacognition, and collective efficacy on team performance. *Journal of Cognitive Engineering and Decision Making*, *11*(4), 382–395.

Hellar, D. B. (2009). *An investigation of data overload in team-based distributed cognition systems.* Unpublished doctoral dissertation, The Pennsylvania State University, University Park, PA.

Hellar, D. B., & McNeese, M. D. (2010). NeoCITIES: A simulated command and control task environment for experimental research. In *Proceedings of the 54th meeting of the Human Factors and Ergonomics Society* (pp. 1027–1031). Los Angeles, CA: SAGE Publications.

Hmelo-Silver, C. E. (2004). Problem-based learning: What and how do students learn? *Educational Psychology Review, 16*(3), 235–266.

Hollan, J., Hutchins, E., & Kirsh, D. (2000). Distributed cognition: Towards a new foundation of HCI. *ACM Transactions on Computer Human Interaction, 7*(2), 174–196.

Hudlicka, E. (2003). To feel or not to feel: The role of affect in human–computer interaction. *International Journal of Human-Computer Studies, 59*(1–2), 1–32.

Hutchins, E. (1995). How a cockpit remembers its speeds. *Cognitive Science, 19*(3), 265–288.

Johnson, M. D., Hollenbeck, J. R., Humphrey, S. E., Ilgen, D. R., Jundt, D. K., & Meyer, C. J. (2006). Cutthroat cooperation: Asymmetrical adaptation to changes in team reward structures. *Academy of Management Journal, 49*, 103–119.

Jones, A. C., & McNeese, M. D. (2006). Cognitive fieldwork in emergency medical services: Developing the I-T-P abstraction hierarchy. In *Proceedings of the 50th annual meeting of the Human Factors and Ergonomics Society* (pp. 501–505). Los Angeles, CA: SAGE Publications.

Jones, R. E. T. (2006). *Studying the impact of fuzzy cognitive map decision aid on team performance and team cognition in a crisis-management, team-based, resource allocation simulation.* Unpublished doctoral dissertation, The Pennsylvania State University, University Park, PA.

Klein, G., Moon, B., & Hoffman, R. R. (2006). Making sense of sensemaking: Alternative perspectives. *Intelligent Systems, 21*(4), 70–73.

Klein, G. A., Orasanu, J., Calderwood, R., & Zsambok, C. E. (Eds.). (1993). *Decision making in action: Models and methods.* Westport, CT: Ablex Publishing.

Kruijff, G. J. M., Colas, F., Svoboda, T., Van Diggelen, J., Balmer, P., Pirri, F., & Worst, R. (2012, March). Designing intelligent robots for human-robot teaming in urban search and rescue. In *AAAI Spring symposium: Designing Intelligent Robots.* Cambridge, MA: AAAI Press.

Mancuso, V. F., Hamilton, K., Tesler, R., Mohammed, S., & McNeese, M. (2013). An experimental evaluation of the effectiveness of endogenous and exogenous fantasy in computer-based simulation training. *International Journal of Games and Computer-Mediated Simulations, 5*, 50–65.

Mancuso, V. F., Minotra, D., Giacobe, N., McNeese, M., & Tyworth, M. (2012, March). idsNETS: An experimental platform to study situation awareness for intrusion detection analysts. In *2012 IEEE international multi-disciplinary conference on Cognitive Methods in Situation Awareness and Decision Support (CogSIMA)* (pp. 73–79). Piscataway, NJ: IEEE.

Marks, M. A., Sabella, M. J., Burke, C. S., & Zaccaro, S. J. (2002). The impact of cross-training on team effectiveness. *Journal of Applied Psychology, 87*(1), 3–13.

Marks, M. A., Zaccaro, S. J., & Mathieu, J. E. (2000). Performance implications of leader briefings and team-interaction Training for team adaptation to novel environments. *Journal of Applied Psychology, 85*(6), 971–986.

Mathieu, J. E., Heffner, T. S., Goodwin, G. F., Cannon-Bowers, J. A., & Salas, E. (2005). Scaling the quality of teammates' mental models: Equifinality and normative comparison. *Journal of Organizational Behavior, 26*, 37–56.

Mathieu, J. E., Heffner, T. S., Goodwin, G. F., Salas, E., & Cannon-Bowers, J. A. (2000). The influence of shared mental models on team process and performance. *Journal of Applied Psychology, 85*, 273–283.

McComb, S., Kennedy, D., Perryman, R., Warner, N., & Letsky, M. (2010). Temporal patterns of mental model converge: Implications for distributed teams interacting in collaboration space. *Human Factors, 52*(2), 264–281.

McGrath, J. E. (1984). *Groups: Interaction and performance.* Englewood Cliffs, NJ: Prentice Hall.

McNeese, M. D. (1992). *Analogical transfer in situated cooperative learning.* Unpublished doctoral dissertation, Vanderbilt University, Nashville, TN.

McNeese, M. D. (1996). An ecological perspective applied to multi-operator systems. In O. Brown & H. L. Hendrick (Eds.), *Human factors in organizational design and management* (Vol. VI, pp. 365–370). The Netherlands: Elsevier.

McNeese, M. D. (2004). How video informs cognitive systems engineering: Making experience count. *Cognition, Technology, and Work, 6*(3), 186–196.

McNeese, M. D., & Ayoub, P. J. (2011). Concept mapping in the design and analysis of cognitive systems: A historical review. In B. M. Moon, R. R. Hoffman, J. D. Novak, & A. J. Cañas (Eds.), *Applied concept mapping: Capturing, analyzing and organizing knowledge* (pp. 47–66). Boca Raton, FL: CRC Taylor & Francis Publishing.

McNeese, M. D., Bains, P., Brewer, I., Brown, C. E., Connors, E. S., Jefferson, T., . . . Terrell, I. S. (2005). The NeoCITIES simulation: Understanding the design and methodology used in a team emergency management simulation. In *Proceedings of the 49th annual meeting of the Human Factors and Ergonomics Society* (pp. 591–594). Los Angeles, CA: SAGE Publications.

McNeese, M. D., Brewer, I., Jones, R. E. T., & Connors, E. S. (2006). Supporting knowledge management in emergency crisis management: Envisioned designs for collaborative work. In J. Yen, & R. L. Popp (Eds.), *Emergent information technologies and enabling policies for counter-terrorism* (pp. 255–280). Hoboken, NJ: Wiley-IEEE Press.

McNeese, M. D., & Brown, C. E. (1986). *Large group displays and team performance: An evaluation and projection of guidelines, research, and technologies.* AAMRL-TR-86-035. Wright-Patterson Air Force Base, OH: Armstrong Aerospace Medical Research Laboratory.

McNeese, M. D., & Forster, P. K. (Eds.). (2017). *Cognitive systems engineering: An integrative living laboratory framework.* Boca Raton, FL: CRC Taylor & Francis Publishing, Inc.

McNeese, M. D., Perusich, K., & Rentsch, J. R. (2000, July). Advancing socio-technical systems design via the living laboratory. In *Proceedings of the Human Factors and Ergonomics Society annual meeting* (Vol. 44, No. 12, pp. 2–610). Los Angeles, CA: SAGE Publications.

McNeese, M. D., Salas, E., & Endsley, M. (Eds.). (2001). *New trends in cooperative activities: System dynamics in complex environments.* Santa Monica, CA: Human Factors and Ergonomics Society.

McNeese, M. D., Theodorou, E., Ferzandi, L., Jefferson, T., Jr., & Ge, X. (2002). Distributed cognition in shared information spaces. In *Proceedings of the 46th annual meeting of the Human Factors and Ergonomics Society* (pp. 556–560). Santa Monica, CA: Human Factors and Ergonomics Society.

McNeese, M. D., Zaff, B. S., & Brown, C. E. (1992). Computer-supported collaborative work: A new agenda for human factors engineering. In *Proceedings of the IEEE National Aerospace and Electronics Conference (NAECON), 2* (pp. 681–686). Dayton, OH: IEEE, Aerospace and Electronic Systems Society.

McNeese, M. D., Zaff, B. S., Citera, M., Brown, C. E., & Whitaker, R. (1995). AKADAM: Eliciting user knowledge to support participatory ergonomics. *International Journal of Industrial Ergonomics, 15*(5), 345–363.

McNeese, N., Pfaff, M., Santoro, G., & McNeese, M. (2008). Team performance in real and virtual worlds: The perceived value of Second Life. In *Proceedings of the 52nd Annual Meeting of the Human Factors and Ergonomic Society* (pp. 1435–1439). New York

City, NY, Human Factors and Ergonomics Society. Sage CA: Los Angeles, CA: SAGE Publications.

McNeese, N. J., Demir, M., Cooke, N. J., & Myers, C. (2018). Teaming with a synthetic teammate: Insights into human-autonomy teaming. *Human Factors*, *60*(2), 262–273.

Miller, D. L., Young, P., Kleinman, D., & Serfaty, D. (1998). *Distributed dynamic decision-making simulation phase I: Release notes and user's manual*. Woburn, MA: Aptima.

Minotra, D. (2012). *The effect of a workload-preview on task-prioritization and task-performance*. Unpublished doctoral dissertation, The Pennsylvania State University, University Park, PA.

Minotra, D., & McNeese, M. D. (2017). Predictive aids can lead to sustained attention decrements in the detection of non-routine critical events in event monitoring. *Cognition, Technology and Work*, *19*(1), 161–177.

Mohammed, S., Hamilton, K., Tesler, R., Mancuso, V., & McNeese, M. (2015). Time for temporal team mental models: Expanding beyond "what" and "how" to incorporate "when". *European Journal of Work and Organizational Psychology*, *24*, 693–709.

Mohammed, S., & Ringseis, E. (2001). Cognitive diversity and consensus in group decision making: The role of inputs, processes, and outcomes. *Organizational Behavior and Human Decision Processes*, *85*, 310–335.

Moon, B., Hoffman, R. R., Novak, J., & Canus, A. (Eds.). (2011). *Applied concept mapping: Capturing, analyzing, and organizing knowledge*. Boca Raton, FL: CRC, Taylor & Francis Publishing.

Moon, H., Hollenbeck, J. R., Humphrey, S. E., Ilgen, D. R., West, B., Ellis, A. P. J., & Porter, C. O. L. H. (2004). Asymmetric adaptability: Dynamic team structures as one-way streets. *Academy of Management Journal*, *47*(5), 681–695.

Nisser, T., & Westin, C. (2006). *Human factors challenges in unmanned aerial vehicles (UAVs): A literature review*. Lund University School of Aviation, Tech Rep. TFHS, Vol. 5.

Norman, D. A. (2013). *The design of everyday things*. New York: Basic Books.

Norman, G. R., & Schmidt, H. G. (1992). The psychological basis of problem-based learning: A review of the evidence. *Academic Medicine*, *67*(9), 557–565.

Patrick, J., James, N., Ahmed, A., & Halliday, P. (2006). Observational assessment of situation awareness, team differences and training implications, *Ergonomics*, *49*(4), 393–417.

Pattipati, K. R., Kleinman, D. L., & Ephrath, A. R. (1983). A dynamic decision model of human task selection performance. *IEEE Transactions on Systems, Man, & Cybernetics*, *SMC-13*(2), 145–166.

Pearsall, M. J., Ellis, A. P. J., & Bell, B. S. (2010). Building the infrastructure: The effects of role identification behaviors on team cognition development and performance. *Journal of Applied Psychology*, *95*, 192–200.

Pfaff, M. S. (2012). Negative affect reduced team awareness: The effects of mood and stress on computer-mediated team communication. *Human Factors*, *54*(4), 560–571.

Pfaff, M. S., & McNeese, M. D. (2010). Effects of mood and stress on team cognition in a simulated task environment. *Theoretical Issues in Ergonomic Science*, *11*(4), 321–339.

Porter, C. O. L. H., Gogus, C. I., & Yu, R. C. (2010). When does teamwork translate into improved team performance? A resource allocation perspective. *Small Group Research*, *41*(2), 221–248.

Randall, K. R., Resick, C. J., & DeChurch, L. A. (2011). Building team adaptive capacity: The roles of sensegiving and team composition. *Journal of Applied Psychology*, *96*, 525–540.

Reichenbach, M., Frederick, T., Cubrich, L., Bircher, W., Bills, N., Morien, M., . . . Oleynikov, D. (2017). Telesurgery with miniature robots to leverage surgical expertise in distributed expeditionary environments. *Military Medicine*, *182*, 316–320.

Reiffers, A. (2010). *Network access control list situation awareness*. Unpublished doctoral dissertation, The Pennsylvania State University, University Park, PA.

Rentsch, J. R., Delise, L. A., Mello, A. M., & Staniewicz, M. J. (2014). Knowledge building in distributed problem-solving teams: The integrative team knowledge building training strategy. *Small Group Research, 45*(5), 568–591.

Rentsch, J. R., Delise, L. A., Salas, E., & Letsky, M. (2010). Facilitating knowledge building in teams: Can a new team training strategy help? *Small Group Research, 41*(5), 505–523.

Resick, C. J., Dickson, M. W., Mitchelson, J. K., Allison, L. K., & Clark, M. A. (2010). Team composition, cognition, and effectiveness: Examining mental model similarity and accuracy. *Group Dynamics: Theory, Research, and Practice, 14*, 174–191.

Resick, C. J., Murase, T., Bedwell, W., Sanz, E., Jimenez, M., & DeChurch, L. A. (2010). Mental model metrics and team adaptability: A multi-facet multi-method examination. *Group Dynamics: Theory, Research, & Practice, 14*, 332–349.

Resick, C. J., Murase, T., Randall, K. R., & DeChurch, L. A. (2014). Information elaboration and team performance: Examining the psychological origins and environmental contingencies. *Organizational Behavior and Human Decision Processes, 124*, 165–176.

Salomon, G. (Ed.). (1993). *Distributed cognitions: Psychological and educational considerations*. Cambridge, UK: Cambridge University Press.

Santoro, G. M. (1995). What is computer-mediated communication? In Z. L. Berge & M. P. Collins (Eds.), Computer mediated communication and the online classroom. Vol. 1: Overview and perspectives (pp. 11–27). Cresskill, NJ: Hampton.

Schmidt, H. G., & Moust, J. H. (2000). Factors affecting small-group tutorial learning: A review of research. In D. Evensen & C. E. Hmelo (Eds.), *Problem-based learning: A research perspective on learning interactions* (pp. 19–52). Hillsdale, NJ: Erlbaum.

Schmidt, K., & Bannon, L. (2013). Constructing computer supported cooperative work: The first quarter century. *Computer Supported Cooperative Work, 22*(4–6), 345–372.

Stachowski, A. A., Kaplan, S. A., & Waller, M. J. (2009). The benefits of flexible team interaction during crisis. *Journal of Applied Psychology, 94*, 1534–1544.

Stasser, G., & Stewart, D. (1992). Discovery of hidden profiles by decision-making groups: Solving a problem versus making a judgment. *Journal of Personality and Social Psychology, 63*(3), 426–434.

Stasser, G., Vaughan, S. I., & Stewart, D. D. (2000). Pooling unshared information: The benefits of knowing how access to information is distributed among group members. *Organizational Behavior and Human Decision Processes, 82*(1), 102–116.

Stout, R. J., Cannon-Bowers, J. A., Salas, E., & Milanovich, D. M. (1999). Planning, shared mental models, and coordinated performance: An empirical link is established. *Human Factors, 41*(1), 61–71.

Stout, R. J., Salas, E., & Carson, R. (1994). Individual task proficiency and team process behavior: What's important for team functioning? *Military Psychology, 6*(3), 177–192.

Tenenberg, J., Roth, W.-F., & Socha, D. (2016). From I-awareness to we-awareness in CSCW. *Computer Supported Cooperative Work, 25*(4–5), 235–278.

Terrell, I. S., McNeese, M. D., & Jefferson Jr., T. (2004). Exploring cognitive work within a 911 dispatch center: Using complementary knowledge elicitation techniques. In *Proceedings of the 48th annual meeting of the Human Factors and Ergonomics Society* (pp. 605–609). Santa Monica, CA: Human Factors and Ergonomics Society.

Tesler, R., Mohammed, S., Hamilton, K., Mancuso, V., & McNeese, M. (2018). Mirror, mirror: The effects of storytelling and guided team reflexivity on team mental models and performance in distributed teams. *Small Group Research, 49*(3), 267–305.

Tiainen, T., & Koivunen, E.-R. (2006). Exploring forms of triangulation to facilitate collaborative research practice: Reflections from a multidisciplinary research group. *Journal of Research Practice*, *2*(2), 1–16.

Tvaryanas, A. P. (2006). *Human factors Considerations in migration of unmanned aircraft system (UAS) operator control*. HSW-PE-BR-TR-2006-0002. Brooke City-Base, TX: 311th Performance Enhancement Directorate.

Volpe, C. E., Cannon-Bowers, J. A., & Salas, E. (1996). The impact of cross-training on team functioning: An empirical investigation. *Human Factors*, *38*(1), 87–100.

Waller, M. J., Gupta, N., & Giambatista, R. C. (2004). Effects of adaptive behaviors and shared mental models on control crew performance. *Management Science*, *50*(4), 1536–1543.

Wellens, A. R. (1986). Use of a psychological distancing model to assess differences in telecommunication media. In L. Parker & C. Olgren (Eds.), *Teleconferencing and electronic media* (Vol. V, pp. 347–361). Madison, WI: Center for Interactive Programs.

Wellens, A. R. (1993). Group situation awareness and distributed decision making: From military to civilian applications. In J. Castellan (Ed.), *Individual and group decision making: Current issues* (pp. 267–291). Hillsdale, NJ: Lawrence Erlbaum Associates.

Whitaker, R. D., Selvaraj, J. A., Brown, C. E., & McNeese, M. D. (1995). *Collaborative design technology: Tools and techniques for improving collaborative design*. AL/CF-TR-1995-0086. Wright-Patterson Air Force Base, OH: Armstrong Aerospace Medical Research Laboratory.

Wohl, J. G. (1981). Force management decision requirements for air force tactical command and control. *IEEE Transactions on Systems, Man, and Cybernetics*, *11*(9), 618–639.

Young, M. F., & McNeese, M. D. (1995). A situated cognition approach to problem solving. In P. Hancock, J. Flach, J. Caird, & K. Vincente (Eds.), *Local applications of the ecological approach to human-machine systems* (pp. 359–391). Hillsdale, NJ: Erlbaum.

Zaff, B. S., McNeese, M. D., & Snyder, D. E. (1993). Capturing multiple perspectives: A user-centered approach to knowledge acquisition. *Knowledge Acquisition*, *5*(1), 79–116.

4 Distributed Cognition in Teams Is Influenced by Type of Task and Nature of Member Interactions

R. Scott Tindale, Jeremy R. Winget, and Verlin B. Hinsz

CONTENTS

The study of behavior in and by groups has a long history in the social and behavioral sciences (Triplett, 1897; Sherif, 1936; Lorge & Solomon, 1955). Due largely to military funding after WWII, many researchers began studying how teams performed in a variety of different contexts and under varying conditions. This work was integrated and summarized in a landmark volume by Steiner (1972). Two of Steiner's main conclusions were that groups rarely perform up to their full potential and group performance was heavily influenced by the type of task on which they worked. These conclusions remain relevant to more recent attempts at theory and research on team performance. Steiner also focused on issues of both coordination and motivation in explaining team performance, which are also still present in current work in this area (cf. Kerr & Hertel, 2011; Rico, Sánchez-Manzanares, Gil, & Gibson, 2008). However, the term "cognition" was rarely if ever used to describe the work Steiner reviewed.

Shortly after Steiner's review, the types of tasks on which groups worked began to change from mostly physical tasks to more cognitive or information processing tasks. In addition, the social and behavioral sciences were being swept up in the cognitive revolution (Lachman, Lachman, & Butterfield, 1979; Newell & Simon, 1972). Individuals (and later groups) began to be seen as information processing systems, and the theories that guided performance research began to focus both on information and the ways in which it was processed. Several review articles on groups published in the 1990s began to reflect this shift in emphasis (Hinsz, Tindale, & Vollrath, 1997; Larson & Christensen, 1993; Thompson & Fine, 1999). Much of the research from this period focused on how information that was distributed among the group members was processed or used by the group (Stasser & Titus, 1985, 1987). Consistent with one of Steiner's (1972) main conclusions, groups rarely performed up to their potential as information processors. However, the focus on information tended to overshadow Steiner's emphases on other concepts like motivation and coordination. Research since the 1990s has begun to reintroduce such notions into models on group information processing (De Dreu, Nijstad, & van Knippenberg, 2008; Abele, Stasser, & Chartier, 2010).

Another recent trend in group research has focused on how simply aggregating individual judgments (sans interaction or communication) can lead to quite accurate group judgments. The power of groups (or more colloquially, the "wisdom of crowds," Surowiecki, 2004) has led to the use of "big data" to help organizations make several different types of decisions (Tetlock & Gardner, 2015). In such instances, group members (broadly defined) serve as data points or information sources but computer algorithms aggregate the information for further processing and final judgments. This has led some researchers to argue member interaction is superfluous, or even detrimental, to team performance (Armstrong, 2006). This would imply using teams to gain information is useful, but the processing of such information should be done elsewhere. However, other research has shown that group member interaction can and should help groups process information in a number of task situations (Mellers et al., 2014; Kerr & Tindale, 2011).

Our goal in the present chapter is to review and integrate research on teams as cognitive or information processing systems (cf. Hinsz, 2001), taking into account how such processes function for, or are affected by, different types of tasks and the nature of member interaction. It will be a targeted review, attempting to highlight key task features and key aspects of member interaction (or lack thereof). One of the key task distinctions we will make draws from Steiner's (1972) distinction between unitary and divisible tasks. Unitary tasks are those where all group members are basically working on the same task together. For example, a team of programmers correcting an error in a computer program would be a unitary task. Divisible tasks are those where each team member (or different subgroups of members) are working on ostensibly different tasks that when combined with the tasks performed by other members (or subgroups) will lead to some collective goal. An example of a divisible task would be an organization launching a new product with some of the team members working on marketing the product, others are working on production, and others still are working on staffing, etc. This distinction is in some senses arbitrary since many unitary tasks could be broken up into subtasks. However, the role of information exchange and interaction is different depending on how easily divisible the task tends to be. We will also discuss how context, task type, and amount of interaction affect how members' knowledge and preferences are combined into final group products (Davis, 1982; Hinsz & Ladbury, 2012). How groups combine their individual preferences, knowledge, skills, etc. to reach groups goals has been an important topic for group performance for both fully interacting and non-interacting (simple aggregate) groups. Some combination processes are possible regardless of levels of interaction (i.e., averaging) but others require greater levels of interaction in order to emerge (Hinsz & Ladbury, 2012; Kerr & Tindale, 2011). Thus, we will emphasize the role of combination processes throughout the chapter. Finally, we will attempt to use current theory and research findings to make suggestions on how best to use teams as information processing systems and to discuss where the field might productively head in the future.

UNITARY TASKS

Simple Aggregation—No Interaction

Although the basic finding has been known since Galton's (1907) wisdom of the crowd, Surowiecki (2004) brought the notion of the wisdom of crowds to the forefront of popular culture. The basic idea behind the wisdom of crowds is that an aggregation of many individual judgments will tend to be more accurate than a randomly selected individual judgment and will often be more accurate than a single judgment from an expert (Steiner, 1972). This phenomenon has been replicated many times in a variety of different judgment task domains (Larrick & Soll, 2006; Surowiecki, 2004). Ariely et al. (2000) showed, assuming pairwise conditional independence and random individual error distributions (although rare in many decision contexts), the average of J probability estimates (J = the number of estimators) will always be better than any of the component individual estimates and as J increases, the average will tend toward perfect calibration diagnosticity (accurate representation of the true state of affairs), even when information provided to the various estimators is less than optimal. In addition, Johnson, Budescu, and Wallsten (2001) empirically showed the accuracy of the average probability estimate to be robust over several conditions, even when individual estimates were not independent. Recent work on forecasting has shown a simple average of multiple independent forecasts will perform better than individual experts and often perform as well as more sophisticated aggregation techniques (Armstrong, 2001).

Larrick and Soll (2006) have explained the advantage of simple averages over individual judgments using the concept of "bracketing." If the group member judgments are independent, different members will make some of the estimates above the "true score" and others below it. Thus, the estimates "bracket" the true score. When this is true, it can be mathematically shown the mean of the multiple estimates will always be more accurate than the average individual judge. If the true score is well bracketed by the multiple estimates (near the median or mean), the aggregate accuracy will be far superior to the typical individual judge. However, even if the true score is closer to one of the tails of the distribution, the mean will still outperform the typical individual, though not to the same degree. Larrick and Soll (2006) also show even when the true score is not bracketed by the estimates, the group (average) will do no worse than the typical individual judge.

From an information processing perspective, bracketing is a function of distributed information. Different group members have different information about the particular judgment context. This differential information access leads to judgments that vary as a function of that information. Assuming the generally available information is not biased toward a particular tail of the distribution, the judgments should randomly vary around the true score. Thus, the wisdom of crowds can be viewed as a function of the natural distribution of information across members.

Although central tendency aggregation models have been shown to do quite well in many situations (Larrick & Soll, 2006), a number of researchers have attempted to improve aggregate forecasts by modifying the aggregation procedure. Budescu and Chen (2014) formulated a method for improving group forecasts by eliminating members whose forecasts detract from the aggregate performance. They had individuals make probabilistic forecasts for a variety of events and then assessed whether the individual's forecast was better or worse when each individual was included in (or removed from) the aggregate. By only including those individuals whose forecasts showed a positive influence on accuracy, they consistently improved the accuracy of the aggregate forecasts relative to the simple average and other less effective weighting schemes, and the improvements persisted for future judgments not used to define the inclusion criteria (see also Mellers et al., 2014). Mannes, Soll, and Larrick (2014) suggest a select-crowd strategy, which ranks judges based on a cue to ability (e.g., the accuracy of several recent judgments) and averages the opinions of the top judges (e.g., the top five). Through both simulation and an analysis of 90 archival data sets, results show select crowds of five knowledgeable judges yield very accurate judgments across a wide range of possible settings—the strategy is both accurate and robust (Mannes et al., 2014). Following this, they examine how people prefer to use information from a crowd. The authors' findings demonstrate people are drawn to experts and dislike crowd averages, but importantly, they view the select-crowd strategy favorably and are willing to use it. The select-crowd strategy is accurate, robust, and appealing as a mechanism for helping individuals tap collective wisdom.

Aggregation with Limited Information Exchange

Although simple aggregation tends to produce fairly accurate decisions, there is little chance for members to share information or defend their positions. In addition, group members often remain unaware of others' positions and the final group product. Although there is evidence that often little is gained by member exchanges of information for some judgment tasks (Armstrong, 2006; Lorenz, Rauhut, Schweitzer, & Helbing, 2011), it is difficult for members with insights or valuable information to have influence without some type of interaction among team members (Kerr & Tindale, 2011). Obviously full group deliberation allows members to share and defend their positions. However, there is evidence the most influential members in freely interacting groups (based on status or confidence) are not always the most accurate or correct (Littlepage, Robison, & Reddington, 1997). Thus, various approaches at compromise procedures have been suggested in which some information exchange is allowed but the procedures minimize conformity pressures and incidental influence.

Probably the most famous of these limited-exchange procedures is the Delphi technique (Dalkey, 1969; Rowe & Wright, 1999, 2001). The Delphi technique has been used frequently for idea generation and forecasting, but it has also been adapted to other situations (Rohrbaugh, 1979). The procedure starts by having a group of (typically) experts make a series of estimates, rankings, idea lists, etc. on some topic of interest to the group or facilitator. The facilitator then compiles the list of member responses and summarizes them in a meaningful way (mean rank or probability estimate, list of ideas with generation frequencies, etc.). The summaries are given back to the group members, and they are allowed to revise their initial estimates. The group members are typically anonymous, and the summaries do not specify which ideas or ratings came from each member. This procedure allows information from the group to be shared among the group members but avoids conformity pressure or undue influence by high-status members. The procedure can be repeated as many times as seems warranted but is usually ended when few if any revisions are recorded. The final outcome can range from a frequency distribution of ideas to a choice for the most preferred outcome or the mean or median estimate. A number of related procedures (e.g., nominal group technique; Van de Ven & Delbecq, 1974) use similar procedures but vary in terms of how much information is shared and whether group members can communicate directly. Overall, the purpose of these procedures is to allow for some information exchange while holding potential distortions due to social influence in check. Research on the Delphi technique has tended to show positive outcomes. Delphi groups do better than single individuals and do at least as well as, if not better than, face-to-face groups (Rohrbaugh, 1979). They have also been found to work well in forecasting situations (Rowe & Wright, 1999, 2001).

A more recent technique is the use of prediction markets (cf. Wolfers & Zitzewitz, 2004). Much like financial markets, prediction markets use buyers' willingness to invest in alternative events (e.g., Great Britain will vote to stay vs. leave the European Union, the U.S. will launch a cyber-attack against Iran in the next year, etc.) as a gauge of their likelihood. They typically do not prohibit direct communication among forecasters/investors/bettors, but in usual practice, there is little, if any, communication. However, because the value placed on the assets is typically set in an open market of buyers and sellers, those already in (or out) of the markets can be informed and swayed by various market indicators (e.g., movements in prices, trading volume, volatility), and thus mutual social influence can occur through such channels. Prediction markets are a dynamic and continuous aggregation process in which bids and offers can be made, accepted, and rejected by multiple parties, and the collective expectations of the "group" can continue to change right up to the occurrence of the event in question (e.g., an election). Except for those with ulterior motives (e.g., to manipulate the market, or to use the market as a form of insurance), investments in such markets are likely to reflect the investors' honest judgments about the relative likelihood of events. Members can use current market values to adjust their thinking and learn from the behavior of other members. However, such investment choices are not accompanied by any explanation or justification. Indeed, such investors may even have incentives to withhold vital information that would make other investors' choices more accurate (e.g., that might inflate the price of a "stock" one wants to accumulate). Thus, in terms of opportunities for mutual education and persuasion, prediction markets fall somewhere between statistical aggregation methods (which allow none) and face-to-face groups (which allow many).

There is now a growing body of evidence supporting of the accuracy of prediction markets (Forsythe, Nelson, Neumann, & Wright, 1992; Rothchild, 2009; Wolfers

& Zitzewitz, 2004). Like much of human judgment, prediction markets sometimes overestimate the likelihood of very rare events (Kahneman &Tversky, 1979), but they have done extremely well at predicting various elections in the U.S. and elsewhere. There is also experimental evidence group members can learn from participating in market-type environments. Maciejovsky and Budescu (2007) had people participate in a competitive auction bidding for information in order to solve the Wason card task, which requires using disconfirming evidence for testing a hypothesis (i.e., overcoming the confirmation bias). Their results showed that participants were better at solving such problems (chose the appropriate evidence in an efficient manner) after having participated in the auctions. Thus, even with very minimal exchange, groups can be very accurate decision makers and their members can gain expertise during the process.

Vroom and Yetton (1973) argued one of the ways managers make decisions is through consultation: The decision is made by the manager but only after getting advice from key members of the team. Vroom and Yetton (1973) argued consultation is optimal when managers do not have all the information at their disposal to make a good decision. Thus, they utilize the information distributed among the rest of the team members. Sniezek and Buckley (1995) referred to this mode of social decision making as the "judge-advisor" systems approach. Such decision systems are quite common in the military and in many organizations (Sniezek et al., 1995). The judge is responsible for the final decision but he/she seeks out suggestions from various advisors. Judge-advisor systems have received a fair amount of research attention (see Bonaccio & Dalal, 2006 for a review). Based on the research just discussed, unless the judge had far more expertise than an advisor, the judge should weight the advice equal to their own opinion. Although receiving advice usually does improve judges' decisions relative to when they receive no advice, a vast amount of research has shown judges tend to weight their own opinions more than twice as much as the advice they receive (Larrick, Mannes, & Soll, 2012). This has been referred to as egocentric advice discounting (Yaniv, 2004; Yaniv & Kleinberger, 2000). This effect has been found to be extremely robust and has been replicated in many decision situations with several types of judges and advisors (Bonaccio & Dalal, 2006).

Judges do take the expertise of the advisors into account when re-evaluating their position. Thus, judges discount less when the advisors are known experts or their past advice has been accurate (Goldsmith & Fitch, 1997). Judges are also more likely to use advice when making judgments in unfamiliar domains (Harvey & Fischer, 1997), and they learn to discount poor advice to a greater degree than good advice (Yaniv & Kleinberger, 2000). However, judges are not always accurate in their appraisals of advisor's expertise. Sniezek and Van Swol (2001) have shown one of the best predictors of judges' use of advice is advisor confidence, which is poorly correlated with advisor accuracy. Discounting occurs less for advice a judge solicits than advice a judge simply receives (Gibbons, Sniezek, & Dalal, 2003). In addition, judges discount less when the task is complex (Schrah, Dalal, & Sniezek, 2006), when there are financial incentives for being accurate (Sniezek & Van Swol, 2001), and when they trust the advisor (Van Swol & Sniezek, 2005). However, discounting is present in virtually all judge-advisor situations, and it almost always reduces decision accuracy.

Much of the research on judge-advisor systems has only allowed advisors to provide judgments or judgments with confidence ratings (Bonaccio & Dalal, 2006). This does not allow judges to hear arguments in support of particular positions or estimates. In addition, most of the judgment tasks used for this research (and for the simple aggregation research discussed previously) are what Laughlin (1980) would call "judgmental," rather than "intellective," tasks. Judgment tasks involve matters of opinion that do not allow group members to actually "demonstrate" the accuracy or correctness of their judgments. On the other hand, intellective tasks are matters of fact for which correct or more accurate responses exist. For intellective tasks, under particular conditions (Laughlin & Ellis, 1986), group members should be able to convince other members their position is correct or most accurate using the information available. However, to do this, members need to be able to discuss and share information relevant to performance on the task. Limited information exchange strategies tend not to allow for such interactions. Kerr and Tindale (2011) showed limited exchange strategies do well when the correct answer is bracketed by the group member preferences or when the correct solution is the most popular. Nevertheless, when correct solutions are only preferred by a minority of the members or are far from the mean or median member positions, such strategies lead to performance much below levels expected for interacting groups. Now, our attention turns toward fully interacting groups.

Fully Interacting Groups

Most of the research on group decision making has focused on groups in which the members meet face-to-face and discuss the decision problem until they reach consensus (Kerr & Tindale, 2004). Early research in this area tended to focus on member preferences as the major feature predicting group decision outcomes (Davis, 1973; Kameda, Tindale, & Davis, 2003). More recent research has focused on how groups process information (Hinsz et al., 1997) and the degree to which the group uses available information (Brodbeck, Kerschreiter, Mojzisch, Frey, & Schulz-Hardt, 2007; Lu, Yuan, & McLeod, 2012). Additionally, the motivational aspects of groups and group members have begun to receive attention (De Dreu et al., 2008). We will focus mainly on the two more recent areas in the sections below.

A popular approach to studying interacting groups working on a unitary task utilizes the hidden profile paradigm (Stasser & Titus, 1985). This paradigm is marked by a biased pattern of information distribution in which, prior to group discussion, some information is common to all group members and other information is unique to individual members. The common or fully shared information favors a suboptimal decision alternative, whereas all the unique information combined reveals the optimal alternative. Ultimately, this "hides" the optimal decision choice from the group as a whole. It can only be discovered when each individual shares their unique information and the group uses this information to inform its decision.

Research on hidden profile tasks has shown groups generally do not exchange information efficiently and decision quality suffers as a result. A meta-analysis of the hidden profile paradigm showed (1) groups mention more pieces of common information than unique information; (2) hidden profile groups are less likely to find the solution than are groups having full information; and (3) information pooling (i.e., percentage of unique information mentioned out of total available information, percentage of unique information out of total discussion) is positively related to decision quality. Moreover, communication medium (i.e., computer-mediated communication vs. face-to-face) does not affect (4) unique information pooling or (5) group decision quality (Lu, Yuan, & McLeod, 2012). However, group size, total information load, the proportion of unique information, task demonstrability, and hidden profile strength (i.e., degree of bias created by the hidden profile) moderated these effects.

Most of the current research findings have been nicely encapsulated by Brodbeck et al. (2007) in their Information Asymmetries Model of group decision making. The model categorizes the various conditions that lead to poor information sharing into three basic categories. The first category, negotiation focus, encompasses the various issues surrounding initial member preferences. If groups view the decision-making task mainly as a negotiation, members negotiating which alternative should be chosen tend to focus on alternatives and not on the information underlying them. The second category, discussion bias, encompasses those aspects of group discussion that tend to favor shared vs. unshared information (e.g., items shared by many members are more likely to be discussed). The third category, evaluation bias, encompasses the various positive perceptions associated with shared information (e.g., shared information is perceived as more valid, sharing shared information leads to positive evaluations by other group members). All three categories are good descriptions of typical group decision making and can lead to biased group decisions and inhibit cross-fertilization of ideas and individual member learning (Brodbeck et al., 2007).

A key aspect of the Information Asymmetries Model is that the various aspects of information processing in typical groups only lead to negative outcomes when information is distributed asymmetrically across group members, as when a hidden profile is present. Although such situations do occur, and groups can make disastrous decisions under such circumstances (Janis, 1982; Messick, 2006), they are not typical of most group decision environments. In situations where members have independently gained their information through experience, the shared information they have is probably highly valid and more useful than unique information or beliefs held by only one member. Thus, the fact members share preferences and information in many group decision contexts is probably adaptive and has generally served human survival well (Hastie & Kameda, 2005; Kameda & Tindale, 2006). In addition, groups are often (but not always) sensitive to cues in the environment that indicate information is not symmetrically distributed (Brauner, Judd, & Jacquelin, 2001; Stewart & Stasser, 1998).

Although minorities often are not very influential in groups, if minority members have at their disposal critical information others do not have and that implies the initial group consensus may be wrong, other group members will pay attention to them. However, such minority effects may only be realized when groups are (or think they are) working on intellective tasks. Several studies have shown moderation effects of tasks having a demonstrably correct solution. Lu and colleagues (2012) found the likelihood of a manifest profile (i.e., all members have access to all information) over a hidden profile group finding the optimal task solution increased when working on tasks with high (vs. low) solution demonstrability (i.e., odds ratios of 15.18 vs. 2.46, respectively). Their results indicate hidden profile tasks without a clear preferred solution are most detrimental to information sharing and decision quality, whereas highly demonstrable tasks increase information sharing (Lu et al., 2012). These findings are consistent with other research showing information pooling is more predictive of decision quality (Mesmer-Magnus & DeChurch, 2009) and group discussions are less likely to focus on common information during high demonstrability tasks (Reimer, Reimer, & Czienskowski, 2010).

Other research provides converging evidence for these claims of the influence of task demonstrability. Specifically, Laughlin, Bonner, and Miner (2002) had 82 four-person cooperative groups and 328 independent individuals solve a random coding of the letters A–J to the numbers 0–9. On each trial the group or individual proposed an equation in letters (e.g., A + D = ?), received the answer in letters (e.g., A + D = B), proposed one specific mapping (e.g., A = 3), received the answer (e.g., True, A = 3), and proposed the full mapping of the ten letters to the ten numbers. Researchers found groups needed fewer trials to find the solution, proposed more complex equations, and identified more letters per equation than each of the best, second-best, third-best, and fourth-best individuals. In this experiment, the nature of the task had a clearly appropriate solution (i.e., it was intellective rather than judgmental), which required demonstrable recognition of correct answers, demonstrable rejection of erroneous answers, and multiple insights into effective collective information processing strategies.

According to the theory of combinations of contributions, the outcomes of group interaction on a task can be predicted by two components: the contributions and the combinations (Hinsz & Ladbury, 2012). The contributions refer to the inputs group members bring with them to the task situation (e.g., cognitive skills, processing goals, etc.). The combinations refer to the aggregation principle by which the contributions are combined to lead to the group outcomes (e.g., strategies to pool, share, and integrate information). Importantly, contributions and combinations directly relate to the cognitive processes involved in how group inputs result in team outcomes on a task (Hinsz & Ladbury, 2012).

Groups always exist in a context, and they are sensitive to this context (Hinsz & Ladbury, 2012). Thus, the combinatorial rule that summarizes the processes by which inputs are transformed into outcomes is dependent on the context as well. One of the key findings concerning how teams process information is the common knowledge effect; that is, information shared by many team members plays a larger role in team process and performance than unshared information (Stasser & Titus, 1985). Given this finding, it seems that to increase the amount of information sharing within a team, all team members should have access to all the information available. Indeed, despite the benefits of such manifest profile groups (e.g., Lu et al., 2012), in such information-rich environments, assigning all information to all members may overload each member's cognitive capabilities.

Tindale and Sheffey (2002) examined ways to optimally assign information to group members. Following a model proposed by Zajonc and Smoke (1959), the researchers assessed the effects of information assignment redundancy and group interaction on group memory performance. Participants in five-person groups received either a full list of consonant-verb-consonant non-word trigrams to memorize or a partial list with each trigram distributed to two group members. Groups recalled trigrams as either coacting or interacting groups. In terms of correct recall, coacting groups outperformed interacting groups, and partial redundancy produced better recall than total redundancy. However, intrusion errors were greatly reduced by group interaction and/or a reduction in the cognitive load on the individual group members (i.e., partial redundancy). Groups in the partial redundancy condition tended to perform near optimal levels. A thought experiment of a similar problem using the ideal group model (Sorkin & Dai, 1994) produced similar results (Wallace & Hinsz, 2010). By comparing distributions of information that were unique to each member of the group, partially redundant among group members, or completely redundant, the simulation indicated partially redundant distributions produced superior memory performance to that of unique or complete redundant conditions.

Motivation in groups has been a topic of interest in social psychology since its earliest days as a field of inquiry (Triplett, 1897). Many studies have focused on how groups affect the amount of effort expended by their members, and both motivation gains and losses have been demonstrated (Kerr & Tindale, 2004; Weber & Hertel, 2007). Motivation has also been an important topic in group, as well as individual, decision making, and until recently the basic motivational assumption was hedonism. Many models of collective decision-making use basic game theoretic, or utility maximization, principles to explain how members both choose initial preferences and move toward consensus (Kahn & Rapoport, 1984). Thus, much of the early work on group decision making tended to treat individual group members as players in a utility maximization game (Budescu, Erev, & Zwick, 1999). Game theory approaches are quite prevalent and also quite useful for understanding social behavior (Kameda & Tindale, 2006), but other motives more associated with the group level of analysis have also been found to be important (Levine & Kerr, 2007). In addition, many of these motivations were discovered because social behavior did not follow game theoretic expectations (Dawes, van de Kragt, & Orbell, 1988).

Probably the most heavily researched of these more recent motives in groups involves the ingroup bias (Hogg & Abrams, 1988). There is now substantial evidence that when group members think about themselves as a group (thus, sharing a social identity), they begin to behave in ways that protect the group from harm or enhance its overall welfare. Many of the implications of this bias are positive for the group, but there are situations where it prevents groups from making good decisions. For example, groups are more likely than individuals to lie about preferences and resources in a negotiation setting (Stawiski, Tindale, & Dykema-Engblade, 2009). Probably the most prominent example in which protecting or enhancing the group's welfare leads to less than optimal decisions is the inter-individual–intergroup discontinuity effect (Wildschut, Pinter, Vevea, Insko, & Schopler, 2003). McCallum et al. (1985) initially demonstrated this effect by comparing individuals to groups when playing a prisoner's dilemma game. The prisoner's dilemma game is a mixed motive game where the dominant, or individually rational, response is not to cooperate with the other player. However, when both players make the non-cooperative choice, they both do poorly. The only collectively rational choice is for both players to cooperate, which leads to the greatest collective payoff and moderate positive gains for each player. When two individuals play the game and can discuss the game before making choices, they both end up cooperating better than 80% of the time. However, when two groups play the game and each group must choose between cooperation and non-cooperation, groups quite often choose not to cooperate. Over multiple plays of the game, groups end up locked in the mutual non-cooperation payoff and earn far worse payoffs compared to the inter-individual situation. This effect has been replicated many times using several types of mixed motive game structures and different sized groups (see Wildschut et al., 2003 for a review).

However, giving groups the right motivation can help groups to be better information processors. De Dreu et al. (2008) developed a model of group judgment and decision making based on the combination of epistemic and social motives. Called the "motivated information processing in groups" model (MIP-G), it argues information processing in groups is better understood by incorporating two somewhat orthogonal motives: high vs. low epistemic motivation and pro-social vs. pro-self motivation. Earlier work on negotiation had shown negotiators that share both high epistemic motivation and a pro-social orientation were better able to find mutually beneficial tradeoffs and reach better integrative agreements as compared to negotiators with any other combination of motives (De Dreu, 2010). Recent research now shows the same appears to hold true for groups working cooperatively to solve a problem or make a decision. According to the model, high epistemic motivation involves a goal to be accurate or correct, which should lead to deeper and more thorough information search and analysis (Kruglanski & Webster, 1996). Work on the information sharing effects has consistently demonstrated instilling a goal of accuracy or defining the task in terms of solving a problem both increase information sharing (Postmes, Spears, & Cihangir, 2001; Stasser & Stewart, 1992). Members high in pro-social motivation help to ensure all types of information held by each member are likely to be disseminated, rather than just information supporting the position held by an individual member. Recent research showing that members focusing on preferences rather than information tends to impede information sharing is quite consistent with this assertion (Mojzisch & Schulz-Hardt, 2010). The model predicts information processing in groups will only approach optimal levels when group members are high on both epistemic motivation and pro-social orientation. This is because it is the only combination that produces both systematic and thorough processing of information in an unbiased manner. The MIP-G model appears to do a good job of explaining several well replicated findings and has fared well in the few direct attempts to test it (Bechtoldt, De Dreu, Nijstad, & Choi, 2010; De Dreu, 2007).

DIVISIBLE TASKS

Steiner's (1972) definition of divisible tasks probably maps more closely onto most organizational team tasks than the unitary tasks discussed thus far. Unfortunately, much more work on information processing has been done on unitary tasks. This is partially a function of using laboratory studies to follow how information flows through a group. Unitary tasks are easier to use when time is limited, and the implications of information are more clearly defined for unitary tasks. However, recent research in organizational contexts has given cognition a much more prominent role (Salas, Goodwin, & Burke, 2009). Another difference between unitary and divisible tasks involves how information is used. For unitary tasks, information exchange and processing are usually oriented toward solving a specific problem or choosing a particular course of action. Perhaps a useful example of how information is processed for divisible tasks is in terms of command and control teams (Cooke, Gorman, Duran, & Taylor, 2007). Each member of the team has a role and responsibilities for the team's performance. Moreover, for many command and control teams, the information is processed so a team decision can be reached.

For divisible tasks such as command and control, information processing can also serve a coordination function. Coordination is one of major obstacles to effective team functioning on divisible tasks. The information needed for performance is distributed among the team members, and communication of this information is required so the team can meet its objectives. Consequently, for divisible tasks, communication is part of the team cognition (Cooke, Gorman, Myers, & Duran, 2013) and is part of the processing of the team's information (Hinsz et al., 1997). Moreover, in a number of command and control situations, subgroups may be working on different aspects of a task that are interdependent. Knowing when other subgroups may complete their task is important for knowing how to judge the timing and performance on their own subgroup (Marks, Mathieu, & Zaccaro, 2001). Thus, information processing is critical for performance on divisible tasks but in different ways (Cooke et al., 2013).

Probably one of the main cognitive constructs relevant to divisible tasks is shared mental models (Cannon-Bowers, Salas, & Converse, 1993; Hinsz, 1995: Mohammed, Ferzandi, & Hamilton, 2010). Mental models refer to mental representations of the task and the behaviors associated with performing the task (Rouse & Morris, 1986). At the team level, mental models also involve roles and interdependencies among team members (Klimoski & Mohammed, 1994; Mohammed et al., 2010). Cannon-Bowers et al. (1993) differentiated between task models and team models, and these were incorporated into the conceptualization of groups as information processors (Hinsz et al., 1997). Task models involve the various steps involved in the task and the resources (equipment, etc.) necessary to accomplish it. Group, or team, models involve the information and skills members have that are relevant to the task and the ways in which their skills and behaviors must be coordinated to move efficiently toward task completion. Such shared cognitive structures help team members coordinate actions and interpret information from other team members in consistent ways. They allow team members to develop similar explanations of the environment and to more effectively communicate implicitly, both of which improve team performance (Rico et al., 2008). Team mental models can enhance performance to the degree the models are accurate and the members all share the same model (Hinsz, 1995; Salas et al., 2009).

A few theorists have noted that a missing component of much team/groups research is time (McGrath & Tschan, 2004; Mohammed, Hamilton, & Lim, 2009). Time or timing is a critical component of divisible tasks where early outcomes inform later decisions and behaviors. Until recently, time has been a missing component of team mental models as well (Mohammed et al., 2009). However, recent research has demonstrated the importance of adding the time dimension to team mental models (Mohammed, Hamilton, Tesler, Mancuso, & McNeese, 2015). They argue that adding the notion of temporal team mental models to the general team mental model framework would improve our understanding of how teams coordinate their actions to achieve better outcomes. Using multiple operationalizations of temporal team mental models, they showed teams that incorporated temporal aspects into their mental models performed better than teams with mental models lacking in such aspects.

Team training on both task and team models tends to improve performance by insuring that all aspects of both models are shared (Cannon-Bowers et al., 1993). A well-known team-training program, Cockpit Resource Management (Weiner, Kanki, & Helmreich, 1993), shows how training helps to create effective mental models. In an attempt to decrease errors in airline cockpit crews, Weiner et al. (1993) had each crewmember cross train on their specific role or task as well as on every other role in the cockpit. This cross-training allowed team members to better understand how their role fit in with other roles and how the information they possessed affected other roles. In addition, team members were trained to feel comfortable communicating the information they had and to argue for its relevance in the presence of higher-status team members. Thus, teams were trained to share both a mental model of the cockpit but also the appropriateness of free-flowing information exchange across team members and status differences. Teams trained in this way showed substantial reductions in errors and an increase in airline safety. Similar performance enhancements have been shown for surgery and other teams in hospitals (King et al., 2008).

However, sharedness for either the task or group model will only enhance performance to the degree the model is accurate (Hinsz, 1995; Hinsz & Ladbury, 2012). Stasser and Augustinova (2008) have shown that hierarchical, distributed decision situations where each member has only incomplete information often produce better outcomes if information is simply sent up through the system by each group member without requiring any type of intermediary judgments by members. However, in practice, many groups assume allowing judgments from various members is useful and use such a model to guide their behavior. Although aggregate judgments by many actors with different types and amount of information tend to be more accurate than judgments made by single individuals (Kerr & Tindale, 2011), in distributed systems where each member has only one type of information, asking all the members to make judgments adds noise to the system. In addition, research has shown it is better for members not to know others might have the same information they do because it reduces their feelings of criticality and decreases the likelihood they will send all their relevant information forward (cf. Kerr & Hertel, 2011). Tschan et al. (2009) have shown critical information easily available to emergency medical teams is often overlooked because each member assumes someone else would have discovered and presented the information if it was relevant. Thus, intuitive mental models shared by group members can inhibit performance if they are inaccurate in terms of the task or if they lead to decreased information sharing.

Another type of cognitive construct that has received a fair amount of attention in the literature is transactive memory (Wegner, 1986; Peltokorpi, 2008). Using an individual-level metaphor, Wegner argued team members encode, store, and retrieve information much like single individuals do (see also Wegner, 1995, for a computer metaphor). However, unlike individuals, teams have multiple information storage units, each associated with a different member. Thus, the memory capacity of a team is considerably larger than that of any given team member. However, for the group to be able to use the additional memory storage efficiently, different team members must encode and store different information. For teams working on divisible tasks, the different aspects of the task often define which member will be responsible for encoding and storing certain types of information. For example, pilots may be responsible for knowing flight plans and schedules, whereas copilots maybe responsible for knowing current protocols for final safety checks. The copilot does not need to remember specific details on the flight plan because he/she could always retrieve them from the pilot, and vice versa. Although groups working on unitary tasks can divide up relevant information about the task and form transactive memory systems (Stewart & Stasser, 1995), such systems tend to form naturally for groups working on divisible tasks (Baumann & Bonner, 2011). Transactive memory systems allow for the efficient storage and retrieval of information and also increase team memory capacity. However, for a transactive memory system to work, the members must share a model of who knows what in that they understand how the information is distributed. Consequently, transactive memory systems are instrumental to the effective functioning of teams processing distributed information with divisible tasks.

Liang, Moreland, and Argote (1995) showed that training groups together as a team, rather than training each individual separately, naturally leads to the formation of more effective transactive memory systems. Three-person teams were trained on how to assemble a small radio. The assembly consisted of three component parts, and each team member was trained on each component. Half of the teams were trained together and practiced the different components as a team. The other half involved each individual member being trained individually and then the three individuals were brought together to work as a team. Teams trained together performed better than teams trained as individuals, and the development of a transactive memory system (which team members were better at and knew more about different components) accounted for the difference. These results have now been replicated a number of times (Moreland, Argote, & Krishnan, 1998; Moreland, 1999) and have generalized to training groups in more natural settings (Peltokorpi, 2008).

Another recent construct beginning to receive research attention is the notion of "macrocognition" (Fiore, Smith-Jentsch, Salas, Warner, & Letsky, 2010). The term, as originally used by McNeese (1986), referred to higher level cognition that would be needed to coordinate human-machine systems, such as AI systems in aircraft that aid pilots in flight. More recent conceptualizations have focused on how teams adapt to complex environments and the new knowledge and cognitive processes that emerge from such adaptations (Fiore et al., 2010). Though a fair amount of research has focused on how team mental models can influence performance, macrocognition research focuses on how team mental models change and become more complex as a function of team performance and adaptation.

Although Steiner (1972) referred to divisible tasks within groups in which different members perform different subtasks, recent theorizing on teams in organizations has conceptualized parts of organizations as multi-team systems (Marks, Mathieu, & Zaccaro, 2001; Zacarro, Marks, & DeChurch, 2011). A multi-team system involves "two or more teams that interface directly and interdependently in response to environmental contingencies toward the accomplishment of collective goals" (Marks et al., 2001, p. 290). Thus, the larger organizational task is divided among different teams. DeChurch and Mathieu (2009) argue such systems can be interdependent in at least three domains: inputs, processes, and outputs. As long as the teams share at least one over-arching goal and show interdependence in at least one domain, the teams can be seen as forming a system within the larger organization.

Many of the same issues associated with teams working on divisible tasks also appear for multi-team systems (Zacarro et al., 2011). Importantly, the two factors that contribute to suboptimal team performance (Steiner, 1972) have also been shown to be important for the motivation (Rico, Hinsz, Burke, & Salas, 2017) and coordination (Rico, Hinsz, Davison, & Salas, 2018) of multi-team systems. Moreover, the motive for ingroup bias would arise for the component teams of a multi-team system such that the other teams in a multi-team system will be perceived as outgroups (Hinsz & Betts, 2011). Consequently, like other forms a team structures, the types of interdependencies among members and teams will influence the nature of the interaction among the teams and their members.

The information processing among the component teams in a multi-team system also has similarities to information processing in teams. The knowledge and information residing within one team may serve as inputs to another team (Hinsz, Wallace, & Ladbury, 2009). Alternatively, the actions of one team are likely to require a certain degree of coordination over time to insure efficient system functioning (Hinsz et al., 2009). A conceptualization of multi-team systems can also serve as a bridge to the recent research focus on science teams (Fiore, 2008). Many wicked societal problems involve issues that span different levels of analysis and scientific disciplines. For example, understanding global warming and finding ways to ameliorate it involves meteorology, chemistry, psychology, and sociology, as well as additional disciplines. Thus, future research that helps to further explain how teams and team systems can operate most efficiently should contribute to the solution of other societal problems as well.

Summary and Conclusions

Research to date shows quite clearly that distributed cognition is one of the strongest influences upon the quality of team performance on a variety of tasks teams confront in organizations. The ability to combine and evaluate information from multiple sources to solve a problem or choose a course of action is what allows teams to perform better than individuals working alone (Hinsz et al., 1997). Recent research on aggregation has shown that even without inter-member communication, the diversity of knowledge across members enhances the accuracy of judgments groups make

(Larrick & Soll, 2006). However, for teams to maximize their potential and to ensure unique information is exchanged, other cognitive and motivational factors must be involved.

First, in addition to unique, distributed knowledge, teams need a core base of shared knowledge so all members can see the relevance of the unique information when it is shared (Laughlin & Ellis, 1986; Cannon-Bowers et al., 1993; Hinsz, 1995). Thus, appropriate shared background knowledge or accurate shared mental models allow distributed cognition to aid in team performance. Such shared cognitions also allow team members to better coordinate their efforts. Second, teams need to have high epistemic and social motivations in order to fully use the information at their disposal (De Dreu et al., 2008). Research has consistently shown viewing tasks as having a correct or optimal solution leads to better information sharing (Stewart & Stasser, 1995). In addition, recent work on team forecasting has found that being open to the opinions of other group members is one of the key predictors to team success and forecasting improvement over time (Mellers et al., 2014). When distributed cognition is yoked with accurate shared knowledge, appropriate motivation, and well-coordinated action, teams should be able to use their distributed knowledge to its fullest extent.

Research on the improved utilization of the distributed cognition in teams has contributed to our general understanding of information processing teams. Moreover, as this chapter illustrates, the research on distributed cognition in teams reifies our classic approaches to group productivity (Steiner, 1972). As the research on multi-team systems reinforces, the types of tasks team members face and the ways in which those tasks are assigned to members have dramatic influences on how team members interact and how they process information. It is also the nature of the interactions among team members, whether they share the information available or keep it to themselves, that impacts how teams distribute and process information. As this chapter illustrates, much has been learned about distributed cognition in teams and there is much more that can be learned by examining factors such as type of task and the nature of member interaction in the study of team performance. More work is also needed on how to design and utilize technology to aid groups in performing cognitive tasks. Many if not most teams now depend on technology to some degree for information storage and/or dissemination. However, as work in AI increases how well technology can learn and adjust from data, teams will probably become more dependent on technology for all aspects of task performance. New research in this area will be key to helping teams develop, use, and learn from distributed cognition environments.

REFERENCES

Abele, S., Stasser, G., & Chartier, C. (2010). Conflict and coordination in the provision of public goods: A conceptual analysis of step-level and continuous games. *Personality and Social Psychology Review, 14*, 385–401.

Ariely, D., Au, W. T., Bender, R. H., Budescu, D. V., Dietz, C. B., et al. (2000). The effect of averaging subjective probability estimates between and within groups. *Journal of Experimental Psychology: Applied, 6*, 130–147.

Armstrong, J. S. (2001). *Principles of forecasting: A handbook for researchers and practitioners*. Boston, MA: Kluwer Academic.

Armstrong, J. S. (2006). Should the forecasting process eliminate face-to-face meetings? *Foresight: The Intervational Journal of Applied Forecasting*, *5*, 3–8.

Baumann, M. R., & Bonner, B. L. (2011). Expected group longevity and expected task difficulty on learning and recall: Implications for the development of transactive memory. *Group Dynamics: Theory, Research and Practice*, *15*(3), 220–232.

Bechtoldt, M. N., De Dreu, C. K. W., Nijstad, B. A., & Choi, H. S. (2010). Motivated information processing, epistemic social tuning, and group creativity. *Journal of Personality and Social Psychology*, *99*, 622–637.

Bonaccio, S., & Dalal, R. S. (2006). Advice taking and decision making: An integrative literature review and implications for the organizational sciences. *Organizational Behavior and Human Decision Processes*, *101*, 127–151.

Brauner, M., Judd, C. M., & Jacquelin, V. (2001). The communication of social stereotypes: The effects of group discussion and information distribution on stereotypic appraisals. *Journal of Personality and Social Psychology*, *81*, 463–471. https://doi.org/10.1037/0022-3514.81.3.463

Brodbeck, F. C., Kerschreiter, R., Mojzisch, A., Frey, D., & Schulz-Hardt, S. (2007). Group decision making under conditions of distributed knowledge: The information asymmetries model. *Academy of Management Journal*, *32*, 459–479.

Budescu, D. V., & Chen, E. (2014). Identifying expertise to extract the wisdom of crowds. *Management Science*, *61*(2), 267–280.

Budescu, D. V., Erev, I., & Zwick, R. (Eds.) (1999). *Games and human behavior.* Mahwah, NJ: Lawrence Erlbaum Associates.

Cannon-Bowers, J. A., Salas, E., & Converse, S. (1993). Shared mental models in expert team decision making. In J. Castellan Jr. (Ed.), *Current issues in individual and group decision making* (pp. 221–246). Hillsdale, NJ: Lawrence Erlbaum.

Cooke, N. J., Gorman, J. C., Duran, J. L., & Taylor, A. R. (2007). Team Cognition in Experienced Command-and-Control Teams. *Journal of Experimental Psychology: Applied, Special Issue on Capturing Expertise across Domains*, *13*, 146–157.

Cooke, N. J., Gorman, J. C., Myers, C. W., & Duran, J. L. (2013). Interactive team cognition. *Cognitive Science*, *37*, 255–285. https://doi.org/10.1111/cogs.12009

Dalkey, N. C. (1969). An experimental study of group opinion. *Futures*, *1*(5), 408–426.

Davis, J. H. (1973). Group decision and social interaction: A theory of social decision schemes. *Psychological Review*, *80*, 97–125.

Davis, J. H. (1982). Social interaction as a combinatorial process in group decision. In H. Brandstatter, J. H. Davis, & G. Stocker-Kreichgauer (Eds.), *Group decision making* (pp. 27–58). London, UK: Academic Press.

Dawes, R. M., van de Kragt, A. J., & Orbell, J. M. (1988). Not me or thee but we: The importance of group identity in eliciting cooperation in dilemma situations. Experimental manipulations. *Acta Psychologica*, *68*, 83–97.

De Dreu, C. K. W. (2007). Cooperative outcome interdependence, task reflexivity, and team effectiveness: A motivated information processing perspective. *Journal of Applied Psychology*, *92*, 628–638.

De Dreu, C. K. W. (2010). Social conflict: The emergence and consequences of struggle and negotiation. In S. T. Fiske, D. T Gilbert, & H. Lindzey (Eds.), *Handbook of social psychology* (5th ed., Vol. 2, pp. 983–1023). New York: Wiley.

De Dreu, C. K. W., Nijstad, B. A., & van Knippenberg, D. (2008). Motivated information processing in group judgment and decision making. *Personality and Social Psychology Review*, *12*, 22–49.

DeChurch, L. A., & Mathieu, J. E. (2009). Thinking in terms of multiteam systems. In *Team effectiveness in complex organizations: Cross-disciplinary perspectives and approaches* (pp. 267–292). New York: Psychology Press.

Fiore, S. M. (2008). Interdisciplinarity as teamwork: How the science of teams can inform team science. *Small Group Research, 39*(3), 251–277.

Fiore, S. M., Smith-Jentsch, K. A., Salas, E., Warner, N., & Letsky, M. (2010). Towards an understanding of macrocognition in teams: Developing and defining complex collaborative processes and products. *Theoretical Issues in Ergonomics Science, 11*(4), 250–271. https://doi.org/10.1080/14639221003729128

Forsythe, R., Nelson, F., Neumann, G. R., & Wright, J. (1992). Anatomy of an experimental political stock market. *American Economic Review, 82*, 1142–1161.

Galton, F. (1907). Vox Populi—The wisdom of the crowd. *Nature, 75*(1949), 450–451.

Gibbons, A. M., Sniezek, J. A., & Dalal, R. S. (2003, November). Antecedents and consequences of unsolicited versus explicitly solicited advice. In D. Budescu (Chair), *Symposium in Honor of Janet Sniezek. Symposium presented at the annual meeting of the society for judgment and decision making*. Vancouver, BC.

Goldsmith, D. J., & Fitch, K. (1997). The normative context of advice as social support. Human *Communication Research, 23*, 454–476.

Harvey, N., & Fischer, I. (1997). Taking advice, accepting help, improving judgment and sharing responsibility. *Organizational Behavior and Human Decision Processes, 70*, 117–133.

Hastie, R., & Kameda, T. (2005). The robust beauty of majority rules in group decisions. *Psychological Review, 112*, 494–508.

Hinsz, V. B. (1995). Mental models of groups as social systems: Considerations of specification and assessment. *Small Group Research, 26*, 200–233.

Hinsz, V. B. (2001). A groups-as-information-processors perspective for technological support of intellectual teamwork. In M. D. McNeese, E. Salas, & M. R. Endsley (Eds.), *New trends in collaborative activities: Understanding system dynamics in complex settings* (pp. 22–45). Santa Monica, CA: Human Factors & Ergonomics Society.

Hinsz, V. B., & Betts, K. R. (2011). Conflict in multiple-team situations. In S. J. Zacarro, M. A. Marks, & L. DeChurch (Eds.), *Multi-team systems* (pp. 289–321). New York: Taylor & Francis.

Hinsz, V. B., & Ladbury, J. L. (2012). Combinations of contributions for sharing cognitions in teams. In E. Salas, S. M. Fiore, & M. P. Letsky (Eds.), *Theories of team cognition: Cross-disciplinary perspectives* (pp. 245–270). New York: Routledge.

Hinsz, V. B., Tindale, R. S., & Vollrath, D. A. (1997). The emerging conception of groups as information processors. *Psychological Bulletin, 121*, 43–64.

Hinsz, V. B., Wallace, D. M., & Ladbury, J. L. (2009). Team performance in dynamic task environments. In G. P. Hodgkinson & J. K. Ford (Eds.), *International review of industrial and organizational psychology* (Vol. 24, pp. 183–216). New York: Wiley.

Hogg, M. A., & Abrams, D. (1988). *Social identification: A social psychology of intergroup relations and group processes*. London, UK: Routledge.

Janis, I. L. (1982). *Groupthink: Psychological studies of policy decisions and fiascoes* (2nd ed.). New York: Houghton Mifflin.

Johnson, T. R., Budescu, D. V., & Wallsten, T. S. (2001). Averaging probability judgments: Monte Carlo analyses of asymptotic diagnostic value. *Journal of Behavioral Decision Making, 14*, 123–140.

Kahn, J. P., & Rapoport, A. (1984). *Theories of coalition formation*. Hillsdale, NJ: Lawrence Erlbaum Associates.

Kahneman, D., & Tversky, A. (1979). Prospect theory: An analysis of decision under risk. *Econometrica, 47*, 263–291.

Kameda, T., & Tindale, R. S. (2006). Groups as adaptive devices: Human docility and group aggregation mechanisms in evolutionary context. In M. Schaller, J. A. Simpson, & D. T.

Kenrick (Eds.), *Evolution and social psychology* (pp. 317–342). New York: Psychology Press.

Kameda, T., Tindale, R. S., & Davis, J. H. (2003). Cognitions, preferences, and social sharedness: Past, present and future directions in group decision making. In S. L. Schneider & J. Shanteau (Eds.), *Emerging perspectives on judgment and decision research* (pp. 458–485). New York: Cambridge University Press.

Kerr, N. L., & Hertel, G. (2011). The Köhler group motivation gain: How to motivate the "weak links" in a group. *Social and Personality Psychology Compass*, *5*(1), 43–55.

Kerr, N. L., & Tindale, R. S. (2004). Small group decision making and performance. *Annual Review of Psychology*, *55*, 623–656.

Kerr, N. L., & Tindale, R. S. (2011). Group-based forecasting: A social psychological analysis. *International Journal of Forecasting*, *27*, 14–40. https://doi.org/10.1016/j.ijforecast.2010.02.001

King, H. B., Battles, J., Baker, D. P. et al. (2008). TeamSTEPPS™: Team strategies and tools to enhance performance and patient safety. In K. Henriksen, J. B. Battles, M. A. Keyes et al. (Eds.), *Advances in patient safety: New directions and alternative approaches* (Vol. 3: Performance and Tools). Rockville, MD: Agency for Healthcare Research and Quality (US). Retrieved from www.ncbi.nlm.nih.gov/books/NBK43686/

Klimoski, R., & Mohammed, S. (1994). Team Mental Model: Construct or metaphor? *Journal of Management*, *20*(2), 403–437. https://doi.org/10.1177/014920639402000206

Kruglanski, A. W., & Webster, D. M. (1996). Motivated closing of the mind: "Seizing" and "freezing". *Psychological Review*, *103*, 263–283.

Lachman, R., Lachman, J. L., & Butterfield, E. C. (1979). *Cognitive psychology and information processing: An introduction*. New York: Psychology Press.

Larrick, R. P., & Soll, J. B. (2006). Intuitions about combining opinions: Misappreciation of the averaging principle. *Management Science*, *52*, 111–127.

Larrick, R. P., Mannes, A. E., & Soll, J. B. (2012). The social psychology of the wisdom of crowds. In J. I. Krueger (Ed.), *Social judgment and decision making* (pp. 227–242). New York: Psychology Press.

Larson, J. R. Jr., & Christensen, C. (1993). Groups as problem-solving units: Towards a new meaning of social cognition. *British Journal of Social Psychology*, *32*, 5–30.

Laughlin, P. R. (1980). Social combination processes in cooperative problem-solving groups on verbal intellective tasks. In M. Fishbein (Ed.), *Progress in social psychology* (pp. 127–155). Hillsdale, NJ: Erlbaum.

Laughlin, P. R., & Ellis, A. L. (1986). Demonstrability and social combination processes on mathematical intellective tasks. *Journal of Experimental Social Psychology*, *22*, 177–189.

Laughlin, P. R., Bonner, B. L., & Miner, A. G. (2002). Groups perform better than the best individuals on letters-to-numbers problems. *Organizational Behavior and Human Decision Processes*, *88*, 605–620.

Levine, J. M., & Kerr, N. L. (2007). Inclusion and exclusion: Implications for group processes. In A. E. Kruglanski & E. T. Higgins (Eds.), *Social psychology: Handbook of basic principles* (2nd ed., pp. 759–784). New York: Guilford Press.

Liang, D. W., Moreland, R. L., & Argote, L. (1995). Group versus individual training and group performance: The mediating role of transactive memory. *Personality and Social Psychology Bulletin*, *21*, 384–393.

Littlepage, G. E., Robison, W., & Reddington, K. (1997). Effects of task experience and group experience on performance, member ability, and recognition of expertise. *Organizational Behavior and Human Decision Processes*, *69*, 133–147.

Lorenz, J., Rauhut, H., Schweitzer, F., & Helbing, D. (2011). How social influence can undermine the wisdom of crowd effect. *Proceedings of the National Academy of Sciences, USA*, *108*, 9020–9025.

Lorge, I., & Solomon, H. (1955). Two models of group behavior in the solution of eureka-type problems. *Psychometrica*, *20*, 139–148.

Lu, L., Yuan, Y., & McLeod, P. L. (2012). Twenty-five years of hidden profile studies: A meta-analysis. *Personality and Social Psychology Review*, *16*, 54–75.

Maciejovsky, B., & Budescu, D. V. (2007). Collective induction without cooperation? Learning and knowledge transfer in cooperative groups and competitive auctions. *Journal of Personality and Social Psychology*, *92*, 854–870. https://doi.org/10.1037/0022–3514.92.5.854

Mannes, A. E., Soll, J. B., & Larrick, R. P. (2014). The wisdom of select crowds. *Journal of Personality and Social Psychology*, *107*(2), 276.

Marks, M. A., Mathieu, J. E., & Zaccaro, S. J. (2001). A temporally based framework and taxonomy of team processes. *Academy of Management Review*, *26*(3), 356–376.

McCallum, D. M., Harring, K., Gilmore, R., Drenan, S., Chase, J., Insko, C. A., et al. (1985). Competition between groups and between individuals. *Journal of Experimental Social Psychology*, *21*, 310–320.

McGrath, J. E., & Tschan, F. (2004). *Temporal matters in social psychology.* Washington, DC: American Psychological Association.

McNeese, M. D. (1986). Humane intelligence: A human factors perspective for developing intelligent cockpits. *IEEE Aerospace and Electronic Systems*, *1*(9), 6–12.

Mellers, B., Ungar, L., Baron, J., Ramos, J., Burcay, B., Fincher, K. et al. (2014). Psychological strategies for winning a geopolitical forecasting tournament. *Psychological Science*, *25*, 1106–1115. https://doi.org/10. 1177/095679761452455

Mesmer-Magnus, J. R., & DeChurch, L. A. (2009). Information sharing and team performance: A meta-analysis. *Journal of Applied Psychology*, *94*(2), 535.

Messick, D. M. (2006). Ethics in groups: The road to hell. In E. Mannix, M. Neale, & A. Ten-Brunsel (Eds.), *Research on managing groups and teams: Ethics in groups* (Vol. 8). Oxford, UK: Elsevier Science Press.

Mohammed, S., Ferzandi, L., & Hamilton, K. (2010). Metaphor no more: A 15-year review of the team mental model construct. *Journal of Management*, *36*(4), 876–910. https://doi.org/10.1177/0149206309356804

Mohammed, S., Hamilton, K., & Lim, A. (2009). The incorporation of time in team research: Past, current, and future. In E. Salas, G. F. Goodwin, & C. S. Burke (Eds.), *Team effectiveness in complex organizations: Cross-disciplinary perspective and approaches* (pp. 321–348). New York: Routledge, Taylor & Francis Group.

Mohammed, S., Hamilton, K., Tesler, R., Mancuso, V., & McNeese, M. (2015). Time for temporal team mental models: Expanding beyond "what" and "how" to incorporate "when". *European Journal of Work and Organizational Psychology*. https://doi.org/10.1080/1359432X.2015.1024664

Mojzisch, A., & Schulz-Hardt, S. (2010). Knowing others' preferences degrades the quality of group decisions. *Journal of Personality and Social Psychology*, *98*, 794–808. https://doi.org/10.1037/a0017627

Moreland, R. L. (1999). Transactive memory: Learning who knows what in work groups and organizations. In L. L. Thompson, J. M. Levine, & D. M. Messick (Eds.), *Shared cognition in organizations: The management of knowledge* (pp. 3–31). Mahwah, NJ: Lawrence Erlbaum Associates Publishers.

Moreland, R. L., Argote, L., & Krishnan, R. (1998). Training people to work in groups. In R. S. Tindale, L. Heath, J. Edwards, E. J. Posavac, F. B. Bryant, Y. Suarez-Balcazar, . . .

J. Myers (Eds.), *Theory and research on small groups* (pp. 37–60). New York: Plenum Press.

Newell, A., & Simon, H. A. (1972). *Human problem solving*. Englewood Cliffs, NJ: Prentice Hall.

Peltokorpi, V. (2008). Transactive memory systems. *Review of General Psychology, 12*, 378–394.

Postmes, T., Spears, R., & Cihangir, S. (2001). Quality of decision making and group norms. *Journal of Personality and Social Psychology, 80*(6), 918.

Reimer, T., Reimer, A., & Czienskowski, U. (2010). Decision-making groups attenuate the discussion bias in favor of shared information: A meta-analysis. *Communication Monographs, 77*(1), 121–142.

Rico, R., Hinsz, V. B., Burke, S., & Salas, E. (2017). A multilevel model of multiteam performance. *Organizational Psychology Review, 7*, 197–226.

Rico, R., Hinsz, V. B., Davison, R. B., & Salas, E. (2018). Structural and temporal influences upon coordination and performance multiteam systems. *Human Resources Management Review, 28*, 332–346.

Rico, R., Sánchez-Manzanares, M., Gil, F., & Gibson, C. (2008). Team implicit coordination processes: A team knowledge-based approach. *Academy of Management Review, 33*, 163–184.

Rohrbaugh, J. (1979). Improving the quality of group judgment: Social judgment analysis and the Delphi technique. *Organizational Behavior and Human Performance, 24*, 73–92.

Rothchild, D. (2009). Forecasting elections: Comparing prediction markets, polls, and their biases. *Public Opinion Quarterly, 73*, 895–916.

Rouse, W. B., & Morris, N. M. (1986). On looking into the black box: Prospects and limits in the search for mental models. *Psychological Bulletin, 100*, 349–363.

Rowe, G., & Wright, G. (1999). The Delphi technique as a forecasting tool: Issues and analysis. *International Journal of Forecasting, 15*, 353–375.

Rowe, G., & Wright, G. (2001). Expert opinions in forecasting: Role of the Delphi technique. In J. S. Armstrong (Ed.), *Principles of forecasting: A handbook of researchers and practitioners* (pp. 125–144). Boston, MA: Kluwer Academic Publishers.

Salas, E., Goodwin, G. F., & Burke, C. S. (2009). *Team effectiveness in complex organizations: Cross-disciplinary perspectives and approaches* (E. Salas, G. F. Goodwin, & C. S. Burke, Eds.). New York: Routledge/Taylor & Francis Group.

Schrah, G. E., Dalal, R. S., & Sniezek, J. A. (2006). No decision-maker is an Island: Integrating expert advice with information acquisition. *Journal of Behavioral Decision Making, 19*(1), 43–60.

Sherif, M. (1936). *The psychology of social norms*. New York: Harper and Brothers.

Sniezek, J. A., & Buckley, T. (1995). Cueing and cognitive conflict in Judge-Advisor decision making. *Organizational Behavior and Human Decision Processes, 62*, 159–174.

Sniezek, J. A., & Van Swol, L. M. (2001). Trust, confidence, and expertise in a judge-advisor system. *Organizational Behavior and Human Decision Processes, 84*(2), 288–307.

Sorkin, R. D., & Dai, H. (1994). Signal detection analysis of the IDEAL group. *Organizational Behavior and Human Decision Processes, 60*, 1–13.

Stasser, G., & Augustinova, M. (2008). Social engineering in distributed decision making teams: some implications for leadership at a distance. In S. Weisband (Ed.), *Leadership at a distance* (pp. 151–167). New York: Lawrence Erlbaum Associates.

Stasser, G., & Stewart, D. (1992). Discovery of hidden profiles by decision-making groups: Solving a problem versus making a judgment. *Journal of Personality and Social Psychology, 63*(3), 426.

Stasser, G., & Titus, W. (1985). Pooling of unshared information in group decision making: Biased information sampling during discussion. *Journal of Personality and Social Psychology*, *48*, 1467–1478.

Stasser, G., & Titus, W. (1987). Effects of information load and percentage of shared information on the dissemination of unshared information during group discussion. *Journal of Personality and Social Psychology*, *53*, 81–93.

Stawiski, S., Tindale, R. S., & Dykema-Engblade, A. (2009). The effects of ethical climate on group and individual level deception in negotiation. *International Journal of Conflict Management*, *20*, 287–308.

Steiner, I. D. (1972). *Group process and productivity*. New York: Academic Press.

Stewart, D. D., & Stasser, G. (1995). Expert role assignment and information sampling during collective recall and decision making. *Journal of Personality and Social Psychology*, *69*, 619–628.

Stewart, D. D., & Stasser, G. (1998). The sampling of critical, unshared information in decision making groups: The role of an informed minority. *European Journal of Social Psychology*, *28*, 95–113.

Surowiecki, J. (2004). *The wisdom of crowds*. New York: Doubleday.

Tetlock, P. E., & Gardner, D. (2015). *Superforcasting: The art and science of prediction*. New York: Crown Publishers.

Thompson, L., & Fine, G. A. (1999). Socially shared cognition, affect, and behavior: A review and integration. *Personality and Social Psychology Review*, *3*, 278–302.

Tindale, R. S., & Sheffey, S. (2002). Shared information, cognitive load, and group memory. *Group Processes and Intergroup Relations*, *5*, 5–18.

Triplett, N. (1897). The dynamogenic factors in pacemaking and competition. *American Journal of Psychology*, *9*, 507–533.

Tschan, F., Semmer, N. K., Gurtner, A., Bizzari, L., Spychiger, M., Breuer, M., & Marsch, S. U. (2009). Explicit reasoning, confirmation bias, and illusory transactive memory: A simulation study of group medical decision making. *Small Group Research*, *40*, 271–300.

Van de Ven, A. H., & Delbecq, A. L. (1974). Nominal vs. interacting group processes for committee decision-making effectiveness. *Academy of Management Journal*, *14*, 203–212.

Van Swol, L. M., & Sniezek, J. A. (2005). Factors affecting the acceptance of expert advice. *British Journal of Social Psychology*, *44*(3), 443–461.

Vroom, V. H., & Yetton, P. (1973). *Leadership and decision-making*. Pittsburgh, PA: University of Pittsburgh Press.

Wallace, D. M., & Hinsz, V. B. (2010). Teams as technology: Applying theory and research to model macrocognition processes in teams. *Theoretical Issues in Ergonomic Science*, *11*, 359–374.

Weber, B., & Hertel, G. (2007). Motivation gains of inferior group members: A meta-analytical review. *Journal of Personality and Social Psychology*, *93*(6), 973–993.

Wegner, D. M. (1995). A computer network model of human transactive memory. *Social Cognition*, *13*, 319–339.

Wegner, D. T. (1986). Transactive memory: A contemporary analysis of the group mind. In B. Mullen & G. R. Goethals (Eds.), *Theories of group behavior* (pp. 185–208). New York: Springer-Verlag.

Weiner, E. L., Kanki, B., & Helmreich, R. L. (1993). *Cockpit resource management*. San Diego, CA: Academic Press.

Wildschut, T., Pinter, B., Vevea, J. L., Insko, C. A., & Schopler, C. A. (2003). Beyond the group mind: A quantitative review of the interindividual-intergroup discontinuity effect. *Psychological Bulletin*, *129*, 698–722

Wolfers, J., & Zitzewitz, E. (2004). Prediction markets. *Journal of Economic Perspectives, 18*(2), 107–126.

Yaniv, I. (2004). Receiving other people's advice: Influence and benefits. *Organizational Behavior and Human Decision Processes, 93*, 1–13.

Yaniv, I., & Kleinberger, E. (2000). Advice taking in decision making: Egocentric discounting and reputation formation. *Organizational Behavior and Human Decision Processes, 83*, 260–281.

Zacarro, S. J., Marks, M. A., & DeChurch, L. (2011). *Multi-team systems*. New York: Taylor & Francis.

Zajonc, R. B., & Smoke, W. (1959). Redundancy in task assignments and group performance. *Psychometrika, 24*, 361–370.

5 Bees Do It
Distributed Cognition and Psychophysical Laws

S. Orestis Palermos

CONTENTS

INTRODUCTION

Within philosophy of mind and cognitive science, the hypothesis of distributed cognition puts forward the provocative assumption that appropriately interacting individuals can give rise to collective entities with cognitive properties, over and above the sum of their individual members' cognitive properties. In such cases, we may speak of a distributed cognitive system or, more loosely, of a group mind (for various formulations of this idea, see Barnier, Sutton, Harris, & Wilson, 2008; Heylighen, Heath, & Van Overwalle, 2004; Hutchins, 1996; Sutton, Harris, Keil, & Barnier, 2010; Sutton, 2008; Theiner, Allen, & Goldstone, 2010; Theiner, 2013a, 2013b; Theiner & O'Connor, 2010; Tollefsen & Dale, 2012; Tollefsen, 2006; Wilson, 2005).

The hypothesis of distributed cognition is a version of a broader hypothesis within philosophy of mind, known as active externalism (Clark & Chalmers, 1998; Menary, 2007; Rowlands, 1999; Wilson, 2000, 2004). Active externalism is the assumption that the material realizers of mind and cognition are not necessarily restricted to the agent's organismic machinery. Instead, mental states and cognitive processes may cross organismic boundaries in a number of ways. One assumption is that cognitive processes (such as memory, reasoning, and perception) may extend to the tools that the organism interacts with in order to perform a cognitive task; this is known as the *hypothesis of extended cognition*. Another assumption is that mental states (such as beliefs and desires) may be partly constituted by aspects of the agent's environment; this is the *extended mind thesis*. Finally, there is also the assumption that mind and cognition may be distributed between several individuals (possibly along with their artifacts); this is the *hypothesis of distributed cognition* (or group minds). Though in this chapter I am mainly interested in the latter hypothesis, the discussion will often involve the other two versions of active externalism too.

In the first instance, active externalism (in all of its forms) is a metaphysical thesis about the nature of mind and cognition. Besides philosophy of mind, however, active externalism is also important from the point of view of philosophy of science and scientific practice itself. Lakatos, the famous philosopher of science, noted that, in the hard core of every scientific research program, there lies a set of fundamental metaphysical assumptions, which provide the program with its distinctive identity (Lakatos, 1970). Seen from this perspective, active externalism has the potential to set the tone for a number of scientific research programs within cognitive science. If cognition can indeed extend beyond the agent's organismic boundaries or even be distributed between several agents at the same time, then cognitive scientists should allow this possibility to guide both the development of theory as well as the design of scientific experiments.

Indeed, a growing body of research within cognitive science appears to be implicitly motivated by or, at least, open to active externalism in the form of the hypothesis of distributed cognition. For example, an increasing volume of studies focuses not only on modeling and understanding swarm intelligence and the collective behavior of animals,[1] but also human collective behaviors, such as sports team performance and interpersonal coordination.[2] Though such studies do not always refer to collective behavior as cognitive behavior, some are open to employing such terminology. As Cooke et al. (2013, p. 256) note, for example:

> The term "cognition" used in the team context refers to cognitive processes or activities that occur at a team level. Like the cognitive processes of individuals, the cognitive processes of teams include learning, planning, reasoning, decision making, problem solving, remembering, designing, and assessing situations [. . .]. Teams are cognitive (dynamical) systems in which cognition emerges through interactions.

Nevertheless, despite a few cognitive scientists' tacit uptake of the hypothesis of distributed cognition, several philosophers and cognitive scientists remain skeptical of it. In an attempt to alleviate their concerns, I explore, in what follows, a worrying objection that may be raised against the view. The objection I have in mind has already been raised against the hypothesis of extended cognition and the extended mind thesis and it centers around their frequent reliance on common-sense functionalism. If common-sense functionalism were to be replaced by the more scientifically informed psycho-functionalism, then those two versions of active externalism would appear untenable, especially from the point of view of scientific practice. As I note, the same objection can be used to target the hypothesis of distributed cognition with equal force. As it happens, however, existing research within cognitive science suggests that the hypothesis of distributed cognition is immune to it. If that's correct, then cognitive scientists (and philosophers alike) should be less disinclined to place active externalism—at least in the form of the hypothesis of distributed cognition—at the core of their research projects.

EXTENDED MIND, EXTENDED COGNITION, AND FUNCTIONALISM

To appreciate what the objection is, it will be useful to see how it has already been raised against the hypothesis of extended cognition and the extended mind thesis. A standard argumentative line for the hypothesis of extended cognition and the extended mind thesis involves two steps and a conclusion.[3]

Standard Argument for Extended Cognition

(1) EXTENDED SYSTEM: Identify a case where organismically internal and external components integrate with each other in an extended system to realize a certain process.
(2) COGNITIVE PROCESS: Demonstrate that the process of the target extended system can be readily accepted as a cognitive process.

CONCLUSION: There exists an extended system (consisting of the integrated internal and external components) that realizes a cognitive process—i.e., there exists an extended cognitive system.

This standard argumentative structure can also be used to produce an argument for the extended mind. The only difference is that, instead of focusing on cognitive processes, the argument for the extended mind focuses on mental states.

Standard Argument for Extended Mind

(1') EXTENDED SYSTEM: Identify a case where organismically internal and external components integrate with each other in an extended system to realize a certain state.
(2') MENTAL STATE: Demonstrate that the state of the target extended system can be readily accepted as a mental state.

CONCLUSION': There exists an extended system (consisting of the integrated internal and external components) that realizes a mental state—i.e., there exists an extended mind.

Part of the above reasoning is implicit in what, within the literature on active externalism, has come to be known as the Parity Principle.

Parity Principle

If, as we confront some task, a part of the world functions as a process which, were it go on in the head, we would have no hesitation in accepting as part of the cognitive process, then that part of the world is (for that time) part of the cognitive process.
(Clark & Chalmers, 1998, p. 8)

Proponents of active externalism offer the Parity Principle as an intuition pump (Clark, 2007; Menary, 2007, 2010). Its main purpose is to remind us that considerations about the constituents of mind and cognition should not be guided by spatial location alone (Clark, 2007). The principle, of course, is not theory free. Essentially, it restates the basic functionalist premise: So long as a process/state realizes a function that we would accept as a specific kind of cognitive/mental function, then our judgments about its cognitive/mental status should not be affected by its material realizers, or—in the case of cognitive and mental extension—where these realizers are located. In other words, the Parity Principle is not a full-fledged argument in itself. It only draws attention to the fact that, in the context of active externalism, and so long as functionalism is true, the location of cognition's material realizers should be of no concern; if functionalism is accepted, then pointing merely to the boundaries of skin and skull in order to deny a process/state cognitive status is question-begging.

This is a good starting point—it prevents one from rejecting the hypothesis of extended cognition and the extended mind thesis from the outset. Nevertheless, to fully accept these two hypotheses, rather than remaining open to them, one should motivate the truth of the premises in the standard arguments just discussed. That is, one should demonstrate that, indeed, there is an overall integrated system that consists of both internal and external components, and that the process, or the state, of the extended system is one that we would readily accept as a cognitive process or as a mental state.

With respect to (1) and (1'), Clark (Clark & Chalmers, 1998; Clark, 2008) has attempted to specify when two systems are integrated by invoking the mathematical concept of a *coupled system* from Dynamical Systems Theory. Following Clark, in Palermos (2014), I have provided an extended analysis of when components form parts of an integrated coupled system by focusing on the underlying mathematical details (Chemero, 2011). Rehearsing the details of this analysis would take us too far afield from the present discussion (a summary is provided later on, in the section entitled "Distributed Cognition and Functionalism," where I discuss the integration of group processes), but the main idea is the following: According to Dynamical Systems Theory, two (or more) systems are integrated, or "coupled," if and only if they bring about a process or a state by mutually interacting with each other on the basis of ongoing feedback loops.[4] To appreciate what this means, taking the existence of ongoing mutual interactions as the criterion of integration leads to the following claims: While a painter and his ladder would not qualify as an integrated system (because there are no ongoing mutual interactions between the two), an agent equipped with a tactile visual substitution system (Tyler, Danilov, & Bach-y-Rita, 2003) or a magnetic perception system (Nagel, Carl, Kringe, Märtin, & König, 2005) would.

Let's now move to the defense of claims (2) and (2'): How do we accept a process or a state—be it wholly internal or extended—as a *cognitive process* or a *mental state*? While the Parity Principle invokes functionalism, it does not specify which kind of functionalism one should focus on. Clark and Chalmers (1998), however, seem to rely on common-sense functionalism, according to which mental states and processes should be categorised and understood in terms of our everyday common-sense folk psychology. This becomes apparent in the discussion of their famous case of Otto and his notebook, which they present as a case for the extended mind thesis. Otto is an Alzheimer's patient who compensates for his failing memory by always carrying a well-organized notebook. In order to claim that Otto believes a piece of information inscribed in his notebook—say that MOMA is on 53rd Street—even before looking it up, Clark and Chalmers (1998) compare him to a normal subject, Inga. Upon hearing about an interesting exhibition, Inga thinks, recalls that the museum is on 53rd Street, and starts walking to the museum. Clark and Chalmers argue that if one wants to say that Inga had her belief before consulting her memory, one must also accept that Otto and his notebook believed the museum was on 53rd Street even before Otto looked up the address in his notebook. This is because the two cases are *functionally on a par*, given our everyday, common-sense understanding of how memory works:

> the notebook plays for Otto the same role that memory plays for Inga; the information in the notebook functions just like the information [stored in Inga's biological memory] constituting an ordinary non-occurrent [i.e., dispositional] belief; it just happens that this information lies beyond the skin.
>
> **(Clark & Chalmers, 1998, p. 13)**

To strengthen their point, Clark and Chalmers spell out the relevant common-sensical intuitions by noting that, judging from the case of biological memory, the availability, portability, and reliability of the resource of information are functionally crucial in determining whether a piece of information can qualify as one's dispositional belief. Specifically, they suggest, in order for any device of information storage to be included into an individual's mind, the following criteria must be met:

> (1) The resource [must] be reliably available and typically invoked.
> (2) Any information thus retrieved [must] be more-or-less automatically endorsed.
> (3) [Any] information contained in the resource should be easily accessible as and when required.
>
> (Clark, 2010, p. 46)[5]

It is apparent, then, that according to Clark and Chalmers, in order to determine whether an extended process (or state) is a cognitive process (or mental state) one must consult their common-sense intuitions. Taking this route to argue for the hypothesis of extended cognition and the extended mind thesis is in fact unsurprising, since, as some authors have noted, these two hypotheses are actually a *consequence* of common-sense functionalism. Weiskopf (2008, p. 267) submits, for example, that "functionalism has all along been committed to the possibility of extrabodily states playing the role of beliefs and desires."

This heavy reliance of the standard arguments on common-sense functionalism creates an easy target for opponents of the hypothesis of extended cognition and the extended mind thesis.[6] One could simply suggest that philosophy of mind and cognitive science should steer away from common-sense functionalism and embrace instead a different kind of functionalism. To see how this strategy works, consider

Rupert (2004) who argues that, if we consider fine-grained functional details, Otto's way of recalling information differs to biological memory to such an extent that the two mechanisms cannot be both treated as mental processes. Specifically, Rupert notes, retrieving information from the notebook does not seem likely to exhibit the "negative transfer" and the "generation" effects, which are typically manifested in the process of recalling information from biological memory.[7] Similarly, Adams and Aizawa (2010, p. 63) note that the extended system of Otto and his notebook is unlikely to exhibit the "primacy" and "recency" effects.[8]

This is indeed a promising strategy for rejecting the claim that Inga and Otto and his notebook are functionally on a par. Essentially, it amounts to giving up common-sense functionalism for what is known as psycho-functionalism. Psycho-functionalism claims that "mental states and processes are just those entities, with just those properties, postulated by the best *scientific* explanation of human behaviour" (Levin, 2018). Contrary to common-sense functionalism, "the information used in the functional characterization of mental states and processes needn't be restricted to what is considered common knowledge or common sense, but can include information available only by careful laboratory observation and experimentation" (Levin, 2018). For instance, Rupert's and Adams and Aizawa's resistance for functionally treating Inga and Otto and his notebook on a par invokes the "negative transfer," "generation," "recency," and "primacy" effects, which are not part of our everyday understanding of how memory works and can only be revealed through careful scientific research.

Taking this line of criticism (as well as drawing on other arguments against the hypothesis of extended cognition and the extended mind thesis) several authors have expressed skepticism towards active externalism, in varying strengths. One may argue, for example, that active externalism is, in principle, incorrect. According to this view, which Rupert (2004, 2009) seems to embrace, cognition is necessarily brain- or at most organism-bound. A weaker form of skepticism is to hold that, as a contingent matter of fact, mind and cognition have, so far, been within the skull; future technological advancements, however, may allow them to extend beyond the brain. This is known as "contingent intracranialism" and it is the position that Adam and Aizawa hold: "Insofar as we are intracranialists, we are what might be called 'contingent intracranialists,' rather than 'necessary intracranialists'" (Adams & Aizawa, 2001, p. 57).

WHICH FUNCTIONALISM?

So far, we have seen that invoking fine-grained details in identifying and categorizing mental states and processes would, necessarily or contingently, speak against the hypothesis of extended cognition and the extended mind thesis. However, this psycho-functionalist approach invites a different problem. By focusing on fine-grained, human-specific details about human psychology, psycho-functionalism is open to the charge of being overly "chauvinistic" (Block, 1980). "Creatures whose internal states share the rough, but not fine-grained, causal patterns of ours wouldn't count as sharing our mental states" (Levin, 2018). Unsurprisingly, it is precisely this problem for psycho-functionalism that Clark puts his finger on in his reply to Rupert's objection against the case of Otto: "just because some alien neural system failed to match our own in various ways (perhaps they fail to exhibit the 'generation effect' during recall [. . .]) we should not *thereby* be forced to count the action of such systems as noncognitive" (Clark, 2008, pp. 114–115).

This appears to be a significant worry against psycho-functionalism, though psycho-functionalists may not be impressed by it. One reason for this, they could argue, is that humans are the only organisms we can be certain of having a rich cognitive and mental life. Since it is the level and kind of human cognition we are primarily interested in, it makes sense that we should set it as our exemplar in our studies and theorizing of cognition in general. Yet this kind of response in favor of psycho-functionalism and against common-sense functionalism may beg the question. For how do we decide whether some behavior is cognitive behavior, in the case of human beings? More crucially, how do cognitive scientists decide what behaviors to focus on, set them apart from behaviors that may not be classed as cognitive (such as merely biological or physiological behaviors), and finally study them in details that, subsequently, could inform psycho-functionalist judgements on what may count as cognitive? Psycho-functionalism seems to be the receiving end of this process, not the starting point.

Perhaps, there is a "mark of the cognitive" (Adams & Aizawa, 2001, 2008) that permeates all cognitive phenomena and which cognitive scientists could use in order to identify and isolate them. But in the actual practice of cognitive science, no such mark of the cognitive exists, and philosophical reflection has made apparent that devising one is particularly elusive, even in theory.[9] So in the absence of a mark of the cognitive, how do cognitive scientists choose what to study?

The answer is that cognitive scientists seem to employ, perhaps only implicitly, common-sense functionalism. They choose to study behaviors that on the basis of common-sense intuitions we would readily classify as cognitive. Of course, one may worry that focusing on behavior to identify cognition sounds too close to the doctrine of analytical behaviorism (Huebner, 2013; Ludwig, 2015). But it should be obvious that this would be an uncharitable characterization of what cognitive scientists do. While cognitive scientists do take as their starting point that some process may be classed as cognitive, because it exhibits behavior that we would normally classify as such, contrary to analytical behaviorism, they do not assume that all there is to cognition is behavior alone.

In fact, this subtle yet important distinction between analytical behaviorism and the practice of employing intelligent behavior as evidence for the presence of cognition is not a novel point. As Graham (2015) notes, Sellars pointed this out a long time ago, by invoking the notion of "attitudinal behaviorism":

> Wilfred Sellars (1912–89), the distinguished philosopher, noted that a person may qualify as a behaviorist, loosely or attitudinally speaking, if they insist on confirming "hypotheses about psychological events in terms of behavioural criteria" (1963, p. 22). A behaviorist, so understood, is someone who demands behavioral evidence for any psychological hypothesis. [. . .] Arguably, there is nothing truly exciting about behaviorism loosely understood. It enthrones behavioral evidence, an arguably inescapable premise in not just psychological science but in ordinary discourse about mind and

> behavior. Just how behavioral evidence should be "enthroned" (especially in science) may be debated. But enthronement itself is not in question. Not so [analytical] behaviorism the doctrine.

Keeping in mind the distinction between "attitudinal" and analytical behaviorism is important, because, in its absence, we would be led to believe that cognitive science embraces, at its starting point at least, analytical behaviorism. In the absence of a mark of the cognitive, cognitive scientists' only way for distinguishing between cognitive and non-cognitive phenomena is by employing their common-sense intuitions about what behavior may count as such—but unlike analytical behaviorism they do not further assume that all there is to cognition is mere behavior. This is what Sellars called "attitudinal" behaviorism, or what we may now refer to as common-sense functionalism.[10] Wilson (2001) summarizes this general methodological attitude within philosophy of mind and cognitive science in the following way:

> In order for something to have a mind, that thing must instantiate at least some psychological processes or abilities. Rather than attempting to offer a definition or analysis of what a psychological or mental process or ability is, let the following incomplete list suffice to fix our ideas: perception, memory, imagination (classical Faculties); attention, motivation, consciousness, decision-making, problem-solving (processes or abilities that are the focus of much contemporary work in the cognitive sciences); and believing, desiring, intending, trying, willing, fearing, and hoping (common, folk psychological states).

Common-sense functionalism therefore appears to play an important role within cognitive science, such that abandoning the view entirely and opting, instead, for psycho-functionalism alone is not as clear-cut a decision to make. Consequently, psycho-functionalist arguments against the hypothesis of extended cognition and the extended mind thesis are, to say the least, inconclusive. Nevertheless, by invoking the principle of charity, in what follows, I will assume that psycho-functionalism is the only correct approach to philosophy of mind and cognitive science. The reason for this is that I wish to consider whether the view, if it were to be retained to the exclusion of common-sense functionalism, could raise equally important problems for active externalism in the form of the hypothesis of distributed cognition.

DISTRIBUTED COGNITION AND FUNCTIONALISM

Arguments for the hypothesis of distributed cognition usually follow a similar structure to standard arguments for the hypothesis of extended cognition and the extended mind thesis.[11]

Standard Argument for Distributed Cognition

(1*) DISTRIBUTED COGNITION: Identify a case where several individuals integrate with each other in a distributed system to realize a certain process.

(2*) COGNITIVE PROCESS: Demonstrate that the process of the target distributed system can be readily accepted as a cognitive process.

C: CONCLUSION*: There exists a distributed system (consisting of all participating individuals) that realizes a cognitive process—i.e., there exists a distributed cognitive system.

The requirement that these two premises be satisfied is evident, for example, in Theiner's (Theiner et al., 2010; Theiner, 2013a) Social Parity Principle, which parallels Clark and Chalmers' original Parity Principle.

Social Parity Principle

If, in confronting some task, a group collectively functions in a process which, were it done in the head, would be accepted as a cognitive process, then that group is performing that cognitive process.

Theiner et al. (2010, p. 384) note that the Social Parity Principle, like the original Parity Principle, is not a full-fledged argument in support of distributed cognition. It acts as an intuition pump for judging whether a collective process may qualify as a collective *cognitive* process, on the basis of functionalism. In this way, the Social Parity Principle can motivate premise (2*) of the standard argument for distributed cognition. That is, it can help judge whether a given collective process can qualify as a cognitive process. Nevertheless, it cannot establish the truth of premise (1*), as it provides no reason for thinking that the target cognitive process is a collective cognitive process. Instead for arguing for this claim, the antecedent of the Social Parity Principle already presupposes it.

In an attempt to ensure that both premise (1*) and premise (2*) are satisfied, I have previously offered the following argument, by drawing on the details of Dynamical Systems Theory (Palermos, 2016).

Argument for Distributed Cognition (ADC)

P1: A process D is brought about on the basis of mutual interactions between the members of a group.

P2: According to the systemic properties and ongoing feedback loops arguments, when component parts mutually interact with each other in order to bring about some process P, there exists (with respect to P) an overall system that consists of all the interacting components at the same time.

C1 (from P1 and P2): With respect to D, the underlying group constitutes a distributed system that consists of all the interacting individuals.

P3: D is a process that, on the basis of common-sense intuitions, we would readily classify as cognitive.

C2 (from C1 and P3): With respect to D, the underlying group constitutes a distributed cognitive system that consists of all the interacting individuals.

In this argument, premises P1, P2, and the conclusion C1 seek to establish the truth of premise (1*) of the standard argument for distributed cognition—i.e., that there is a process (i.e., systemic behavior), which is generated by an integrated distributed system. While this is not the place to rehearse the details of the systemic properties and ongoing feedback loops arguments mentioned in P2, it is worth mentioning the main ideas behind them.

The systemic properties argument points out that, according to Dynamical Systems Theory, when individuals engage in ongoing mutual interactions, there arise novel systemic properties in the form of unprecedented regularities of behavior. These properties do not belong to any of the contributing members (or their aggregate), but to their coupled system as a whole. Accordingly, to account for such systemic properties, we have to postulate the overall distributed system. (Alternatively, distributed systems cannot be ontologically eliminated, because they are necessary for accounting for certain systemic properties.)

The ongoing feedback loops argument holds that, according to Dynamical Systems Theory, when individuals mutually interact—on the basis of ongoing feedback loops—to bring about some behavior, they form a nonlinear, causal amalgam that cannot be decomposed in terms of distinct inputs and outputs from the one agent to the other. The reason is that, because of the ongoing feedback loops, the way each individual affects others is not entirely endogenous to itself and, conversely, the way it is affected by others is not entirely exogenous either. Accordingly, to account for the way individuals behave, we cannot but postulate the overall distributed system they form part of.

This suggests that when individuals bring about a behavior by mutually interacting with each other, there are good reasons for postulating an overall distributed system that consists of all of them. In other words, there are good reasons for thinking that premise (1*) of the standard argument for distributed cognition is satisfied. Nevertheless, in order for the hypothesis of distributed cognition to be satisfied, it is also necessary to demonstrate that the relevant distributed system is a distributed *cognitive* system. This further step requires that premise (2*) of the standard argument for distributed cognition be true, which requires showing that the behavior of the distributed system is *cognitive* behavior. In ADC, this is established on the basis of premise P3, which invokes common-sense functionalism.

This reliance of premise P3 on common-sense functionalism means that psycho-functionalism can raise problems for the hypothesis of distributed cognition as this is motivated by ADC. Moreover, even if, unlike ADC, the Social Parity Principle or the standard argument for distributed cognition invoked psycho-functionalism, instead of common-sense functionalism, but there existed no distributed behavior that could be classed as cognitive behavior by the standards of our best scientific theories of cognition, then the prospects of vindicating the hypothesis of distributed cognition would appear bleak.

To counter this worry, in the following section, I review a recent study on the way bee colonies reach decisions on new nest locations. The study demonstrates that even if the Social Parity Principle, premise (2*) of the standard argument for distributed cognition, and premise P3 of ADC invoked psycho-functionalism (instead of common-sense functionalism), the decision-making process of bee colonies would still qualify as a case of distributed cognition.

BEES DO IT

Colonies of the European honey bee (*Apis Mellifera*) reproduce via fission. The old queen flees the nest taking with her thousands of scout bees. During this event, part of the swarm engages in a decision-making process to identify the best location for building a new nest. Scout bees explore the environment, and upon locating a potential nest location, they rejoin the swarm in order to recruit other bees to that location through the waggle dance. When a bee that is committed to a potential nest site meets a scout that propounds an alternative site, she may receive a stop signal. Bees that receive several stop signals fall back to an uncommitted state. This is an ongoing process, during which bees continuously affect each other on the basis of positive (recruitment signals) and negative (stop signals) feedback loops. Eventually, the process ends with a decision being made at the colony level, when bees committed to the same nest site reach a quorum.[12]

In a recent article, Reina, Bose, Trianni, and Marshall (2018) report that they have run computer-based stochastic simulations of a model representing the way the bee colony reaches the above decision. Their nonlinear dynamical model was empirically derived from previous field observations. Astonishingly, their simulations demonstrate that the superorganism's (i.e., the bee colony's) decision-making process manifests the same psychophysical laws—Weber's Law, Hick-Hyman's Law and Piéron's Law—that human brains obey when they engage in decision making.

Weber's Law states that the brain is able to detect the option of the best quality when the difference between the values of the qualities selected from is above a minimum value, and that this required minimum value increases as the values of the compared qualities increase. That is, there is a linear relationship between the values of the qualities selected from and the value of the minimum difference between the two options that is required for successful selection. For example, the human brain can tell whether a set of 30 dots is larger than a set of 25 dots, but it has problems when trying to reach a decision between a set of 150 and 155 dots, even though, in both cases, the difference is 5 dots. In Reina et al.'s (2018) study, the model of the bee colony manifested the same linear dependence between the required minimum quality difference between options (i.e., nest locations) and their mean quality. Piéron's Law states that when the two options are of high quality, the brain reaches decisions faster compared to when the two options are of lower quality. Reina et al.'s study found that the honeybee colony model is faster to decide between two nest locations of high quality compared to two nest locations of low quality. Hick-Hyman's Law suggests that the time the brain requires to reach a decision increases as the number of options increases. In line with this law, Reina et al. found that the model of the bee colony took longer to reach decisions when the amount of alternative nest locations increased.[13]

Drawing on these results the authors conclude that the bee colony, as a distributed dynamical system, obeys the same psychophysical laws as the human brain. In their words:

> This study shows for the first time that groups of individuals, in our case honeybee colonies, considered as a single superorganism, might also be able to obey the same

> laws. Similarly to neurons, no individual explicitly encodes in its *simple* actions the dynamics determining the psychophysical laws; instead it is the group as a whole that displays such dynamics.
>
> **(Reina et al., 2018, p. 5)**

A few points are in order here. First, the fact that the bees mutually interact with each other on the basis of ongoing positive (recruitment) and negative (inhibitory) feedback loops in order to reach a decision indicates that P1 of ADC is satisfied. Second, the fact that Reina et al. employ a nonlinear dynamical model suggests that their experiment is consistent with P2 of ADC, which presupposes the validity of two arguments derived from Dynamical Systems Theory. From this, C3 of ADC follows, which affirms that there exists an integrated distributed system—i.e., what Reina et al. refer to as the "superorganism." Reina et al.'s study, therefore, satisfies premise (1*) of the standard argument for distributed cognition in precisely the way ADC suggests.

What of premise (2*)? Premise (2*) is meant to establish that the relevant distributed system manifests behavior that can be classed as *cognitive* behavior. ADC seeks to affirm this with P3, by invoking common-sense functionalism. As noted in the previous section, this is going to be problematic if we only retain psycho-functionalism—especially if there are no collective processes that would qualify as cognitive processes by psycho-functionalist standards. Nevertheless, Reina et al.'s experiment focuses on the cognitive process of decision making, not as this is understood on the basis of common-sense functionalism, but instead as it is perceived through the lens of psycho-functionalism. Their results demonstrate that the decision-making behavior of the bee colony obeys the same laws as human decision making, when the latter is studied on the basis of careful laboratory observation and experimentation.

This suggests that even if one were to modify ADC such that P3 invoked psycho-functionalism, instead of common-sense functionalism, Reina et al.'s study would still carry ADC to its final conclusion, C2: With respect to the bee colony's decision-making process, there exists a distributed system—i.e., the "superorganism"—that is a distributed *cognitive* system.

Reina et al.'s study therefore indicates that, in the case of the bee colony's decision-making process, all premises of both the standard argument for distributed cognition and (the modified version of) ADC can be satisfied, even by psycho-functionalist standards. In result, there is at least one case for thinking that psycho-functionalism cannot be used against the hypothesis of distributed cognition.

DISCUSSION: PSYCHO-FUNCTIONALISM AND ACTIVE EXTERNALISM

To reiterate, psycho-functionalist opponents of active externalism argue against the view by appealing to psychophysical laws discovered by the best scientific theories of human cognition and psychology:

> Insofar as such laws govern processes in the core of the brain, but not combinations of brains, bodies, and environments, there are principled differences between intracranial brain processes and processes that span brains, bodies, and environments.
>
> **(Adams & Aizawa, 2010, p. 61)**

Reina et al.'s study, however, indicates that, even by the psycho-functionalist's standards, there is a good case to be made against not only necessary intracranialism but also contingent intracranialism.

One possible objection to this is to claim that it does not present us with a case of *human* distributed cognition, but with a case of swarm cognition. Such an objection, however, would be missing the point. If the letter of functionalism (be it common-sense or psycho-functionalism) is to be followed, this should make no difference at all. According to functionalism, in all of its forms, one should not focus on the properties of the supervenient base. The fact that the underlying components of the distributed cognitive system are bees instead of humans should be irrelevant, especially when the psychophysical laws obeyed by the target system are revealed by scientific studies of the behavior of the human brain. Reina et al.'s experiment demonstrates that the bee colony's process of decision making is *human-like*. Psycho-functionalists must therefore accept that this is a genuine case of distributed cognition, even if the underlying group is made up of bees. One may then add that given humans possess more complex and advanced means for communication and interaction, it is only a matter of time before we discover that distributed systems made up of humans obey (or, with advancements in communication technologies, will obey) the requisite psychophysical laws. Cognitive scientists must simply remain open-minded and keep their active externalist goggles on.[14]

But if active externalism in the form of the hypothesis of distributed cognition is by psycho-functionalist standards true, does this mean that it is also true in the form of the hypothesis of extended cognition and the extended mind thesis? Contentiously, one could argue that the hypothesis of distributed cognition is more radical than the hypothesis of extended cognition and the extended mind thesis, because of how widely cognition is dispersed in the former case. Accordingly, if the more radical hypothesis is true, then we should not be surprised if the more moderate versions of active externalism are also true—indeed we might expect them to be so. For example, one can imagine that, even though, currently, there are no extended systems that manifest the effects that Adams, Aizawa, and Rupert point to, in the future, there could be brain-machine interfaces that would augment individuals' memory systems in a way that would obey the same psychophysical laws that biological memory does.

If we follow this line of thought, however, a number of questions arise: Why build into such memory extensions the downsides of biological memory as these are captured by some psychophysical laws? Why, for example, would we like to incorporate the negative transfer, primacy, and recency effects? And if we only incorporated the positive characteristics of biological memory, how many do we need for the extended system to qualify, by the lights of psycho-functionalism, as a cognitive system? Conversely, if a human subject does not exhibit all of the negative transfer, primacy, or recency effects, does their system for information encoding, storage, and retrieval fail to qualify as a memory system? As far as I can see, no straightforward answer exists to these puzzles, because behind them lies a more fundamental question: Which are the *essential* psychophysical laws that a process must manifest in order to qualify as a process of a specific cognitive kind, besides the coarse-grained characterization that common-sense functionalism has to offer for it?

Along with the other objections mentioned in the section entitled "Which Functionalism?," these are important questions that psycho-functionalism needs to address before it can convincingly weight against active externalism in any of its forms. Nevertheless, even if proponents of psycho-functionalism did come up with solutions to all of these problems, the foregoing suggests that they should remain open to active externalism, at least in the form of the hypothesis of distributed cognition.

This leaves us with two general points. First, from the point of view of philosophy of mind, psycho-functionalism does not, in principle, commit one to either necessary or contingent intracranialism. Second and relatedly: From the point of view of philosophy of science and scientific practice itself, cognitive scientists should be less hesitant to incorporate the hypothesis of distributed cognition in the hard cores of their research programs, whether these are further guided by either psycho- or common-sense functionalism.

NOTES

1. See, for example (Obuko, 1986; Niwa, 1994; Parunak, 1997; Li, Yang, & Peng, 2009; Becco, Vandewalle, Delcourt, & Poncin, 2006; Peng, Li, Yang & Liu, 2010; Turnstrøm et al., 2013; Li, Peng, Kurths, Yang, & Schellnhuber, 2014; Attanasi et al., 2015).
2. See, for example (Schmidt, Bienvenu, Fitzpatrick, & Amazeen, 1998; Riley, Richardson, Shockley, & Ramenzoni, 2011; Marsh et al., 2009; Schmidt & Richardson, 2008; Duarte et al., 2013a, 2013b; Dale, Fusaroli, Duran, & Richardson, 2013; Richardson, Dale, & Marsh, 2014).
3. A number of alternative arguments for the hypothesis of extended cognition exist. See, for example, Menary (2006; 2007); Heersmink (2015).
4. A state can be thought of as a process that is at equilibrium.
5. This paper has been available online since 2006. These criteria, however, date even earlier as they had already made their appearance in Clark & Chalmers (1998) (although the phrasing was somewhat different). Also, in Clark & Chalmers (1998, p. 17), the authors consider a further criterion: "Fourth, the information in the notebook has been consciously endorsed at some point in the past, and indeed is there as a consequence of this endorsement." As the authors further note, however, "the status of the fourth feature as a criterion for belief is arguable (perhaps one can acquire beliefs through subliminal perception, or through memory tampering?)," so they subsequently drop it.
6. A further problem for the extended mind thesis as motivated by the case of Otto and his notebook is that the two of them may not actually qualify as a coupled system, as this is understood within Dynamical Systems Theory. In Palermos (2014), I note that this is not necessarily an unwelcome result since, intuitively, it is not so obvious that Otto and his notebook form an extended mind. Nevertheless, Dynamical Systems Theory can still motivate the hypothesis of extended cognition on the basis of more plausible examples, such as agents densely interacting with tactile visual substitution systems and magnetic perception systems (for more details, see Palermos, 2014).
7. According to the 'negative transfer' effect, previous learning negatively affects subjects' ability to learn and recall new associations. According to the 'generation' effect, subjects who generate their own meaningful associations between stored information are better at recalling it. For more details see Rupert (2004).
8. For example, in a free recall task, where subjects are asked to memorize a list of 20 words, the words at the beginning and at the end of the list are more likely to be correctly recalled than the rest. This is supposed to be the consequence of the primacy and recency effects on biological memory, respectively.
9. For details on the debate on the mark of the cognitive, how it may be used against active externalism, and the considerable difficulty to come up with an unproblematic account

for such a concept, see Clark (2010), Menary (2006), Adams and Aizawa (2001, 2008, 2010), Ross and Ladyman (2010).

10. According to common-sense functionalism, behavior is a significant (though not the only) part of cognitive phenomena. Pain, for example, may be broadly characterized as "the state that tends to be caused by bodily injury, to produce the belief that something is wrong with the body and the desire to be out of that state, to produce anxiety, and, in the absence of any stronger, conflicting desires, to cause wincing or moaning" (Levin, 2018).
11. Besides standard arguments, there are a number of ways that one could argue for the hypothesis of distributed cognition. For a comprehensive list of the existing approaches, see Theiner (2015).
12 For more details, see Reina et al. (2018).
13. See "Honeybees may unlock the secrets of how the human brain works" (March 27, 2018), https://medicalxpress.com/news/2018-03-honeybees-secrets-human-brain.html (retrieved 17 December 2019).
14. A research program that accepts active externalism could also be beneficial for understanding the brain. As Reina et al. (2018, p. 5) note, "research synergies between neuroscience and collective intelligence studies can highlight analogies that could help better to understand both systems."

REFERENCES

Adams, F., & Aizawa, K. (2001). The bounds of cognition. *Philosophical Psychology*, *14*(1), 43–64.

Adams, F., & Aizawa, K. (2008). *The bounds of cognition*. Oxford: Blackwell Publishing Ltd.

Adams, F., & Aizawa, K. (2010). Defending the bounds of cognition. In R. Menary (Ed.), *The extended mind*. Cambridge, MA: MIT Press.

Attanasi, A., Cavagna, A., Del Castello, L., Giardina, I., Jelic, A., Melillo, S., et al. (2015). Emergence of collective changes in travel direction of starling flocks from individual birds' fluctuations. *Journal of the Royal Society, Interface*, *12*(108), 20150319.

Barnier, A. J., Sutton, J., Harris, C. B., & Wilson, R. A. (2008). A conceptual and empirical framework for the social distribution of cognition: The case of memory. *Cognitive Systems Research*, *9*(1–2), 33–51. https://doi.org/10.1016/j.cogsys.2007.07.002.

Becco, C., Vandewalle, N., Delcourt, J., & Poncin, P. (2006). Experimental evidences of a structural and dynamical transition in fish school. *Physica A: Statistical Mechanics and its Applications*, *367*, 487–493.

Block, N. (1980). Troubles with functionalism. In *Readings in the philosophy of psychology* (Vols. 1 and 2). Cambridge, MA: Harvard University Press.

Chemero, A. (2011). *Radical embodied cognitive science*. MIT press.

Clark, A. (2007). Curing cognitive hiccups: A defense of the extended mind. *The Journal of Philosophy*, *104*, 163–192.

Clark, A. (2008). *Supersizing the mind: Embodiment, action, and cognitive extension*. New York: Oxford University Press.

Clark, A. (2010). Memento's revenge: The extended mind, extended. In R. Menary (Ed.), *Extended mind*. Cambridge, MA: MIT Press.

Clark, A., & Chalmers, D. (1998). The extended mind. *Analysis*, *58*, 10–23.

Cooke, N. J., Gorman, J. C., Myers, C. W., & Duran, J. L. (2013). Interactive team cognition. *Cognitive Science*, *37*(2), 255–285.

Dale, R., Fusaroli, R., Duran, N., & Richardson, D. C. (2013). The self-organization of human interaction. *Psychology of Learning and Motivation*, *59*, 43–95.

Duarte, R., Araujo, D., Correia, V., Davids, K., Marques, P., & Richardson, M. J. (2013a). Competing together: Assessing the dynamics of team–team and player–team synchrony in professional association football. *Human Movement Science*, *32*(4), 555–566.

Duarte, R., Araujo, D., Folgado, H., Esteves, P., Marques, P., & Davids, K. (2013b). Capturing complex, non-linear team behaviours during competitive football performance. *Journal of Systems Science and Complexity*, *26*(1), 62–72.

Graham, G. (2015). Behaviorism. In E. N. Zalta (Ed.), *The Stanford encyclopedia of philosophy* (Spring 2015 ed.). Retrieved from http://plato.stanford.edu/archives/spr2015/entries/behaviorism

Heersmink, R. (2015). Dimensions of integration in embedded and extended cognitive systems. *Phenomenology and the Cognitive Sciences*, *14*, 577–598.

Heylighen, F., Heath, M., & Van Overwalle, F. (2004). The Emergence of Distributed cognition: A conceptual framework. In *Proceedings of collective intentionality IV.*

Huebner, B. (2013). *Macrocognition: Distributed minds and collective intentionality.* New York: Oxford University Press.

Hutchins, E. (1996). *Cognition in the wild.* Cambridge, MA: MIT Press.

Lakatos, I. (1970). Falsification and the methodology of scientific research programmes. In I. Lakatos & A. Musgrave (Ed.), *Criticism and the growth of knowledge.* Cambridge: Cambridge University Press.

Levin, J. (2018, Fall). Functionalism. In Edward N. Zalta (Ed.), *The Stanford Encyclopedia of Philosophy.* Retrieved from https://plato.stanford.edu/archives/fall2018/entries/functionalism/.

Li, L., Peng, H., Kurths, J., Yang, Y., & Schellnhuber, H. J. (2014). Chaos-order transition in foraging behavior of ants. *Proceedings of the National Academy of Sciences*, *111*(23) 8392–8397. https://doi.org/10.1073/pnas.1407083111.

Li, L., Yang, Y., & Peng, H. (2009). Fuzzy system identification via chaotic ant swarm. *Chaos, Solitons & Fractals*, *41*(1), 401–409.

Ludwig, K. (2015). Is distributed cognition group level cognition? *Journal of Social Ontology*, *1*(2), 189–224.

Marsh, K. L., Richardson, M. J., & Schmidt, R. C. (2009). Social connection through joint action and interpersonal coordination. *Topics in Cognitive Science*, *1*(2), 320–339.

Menary, R. (2006). Attacking the bounds of cognition. *Philosophical Psychology*, *19*(3), 329–344.

Menary, R. (2007). *Cognitive integration: Mind and cognition unbound.* London: Palgrave McMillan.

Menary, R. (2010). Introduction: The extended mind in focus. In R. Menary (Ed.), *The extended mind.* Cambridge, MA: MIT Press.

Nagel, S. K., Carl, C., Kringe, T., Märtin, R., & König, P. (2005). Beyond sensory substitution—Learning the sixth sense. *Journal of Neural Engineering*, *2*(4), R13—R26.

Niwa, H. S. (1994). Self-organizing dynamic model of fish schooling. *Journal of Theoretical Biology*, *171*(2), 123–136.

Obuko, A. (1986). Dynamical aspects of animal grouping: Swarms, schools, flocks, and herds. *Advances in Biophysics*, *22*, 1–94.

Palermos, S. O. (2014). Loops, constitution, and cognitive extension. *Cognitive Systems Research*, *27*, 25–41.

Palermos, S. O. (2016). The dynamics of group cognition. *Minds and Machines*. *26*(4), 409–440.

Parunak, H. V. D. (1997). "Go to the ant": Engineering principles from natural multi-agent systems. *Annals of Operations Research*, *75*, 69–101.

Peng, H., Li, L., Yang, Y., & Liu, F. (2010). Parameter estimation of dynamical systems via a chaotic ant swarm. *Physical Review E*, *81*(1), 016207.

Reina, A., Bose, T., Trianni, V., & Marshall, J. A. (2018). Psychophysical Laws and the Superorganism. *Scientific Reports*, *8*(1), 4387.

Richardson, M. J., Dale, R., & Marsh, K. L. (2014). Complex dynamical systems in social and personality psychology: Theory, modeling, and analysis. In *Handbook of research methods in social and personality psychology* (pp. 253–282). Cambridge: Cambridge University Press.

Riley, M. A., Richardson, M. J., Shockley, K., & Ramenzoni, V. C. (2011). Interpersonal synergies. *Frontiers in Psychology*, *2*, 38.

Ross, D., & Ladyman, J. (2010). The alleged coupling-constitution fallacy and the mature sciences. In Menary (Ed.), *The extended mind*. Cambridge, MA: MIT Press.

Rowlands, M. (1999). *The body in mind: Understanding cognitive processes*. New York: Cambridge University Press.

Rupert, R. D. (2004). Challenges to the hypothesis of extended cognition. *Journal of Philosophy*, *101*, 389–428.

Rupert, R. D. (2009). *Cognitive systems and the extended mind*. Oxford: Oxford University Press.

Schmidt, R. C., Bienvenu, M., Fitzpatrick, P. A., & Amazeen, P. G. (1998). A comparison of intra-and interpersonal interlimb coordination: Coordination breakdowns and coupling strength. *Journal of Experimental Psychology: Human Perception and Performance*, *24*(3), 884.

Schmidt, R. C., & Richardson, M. J. (2008). Dynamics of interpersonal coordination. In *Coordination: Neural, behavioral and social dynamics* (pp. 281–308). Berlin: Springer.

Sellars, W. (1963). Philosophy and the scientific image of man. In *Science, perception, and reality* (pp. 1–40). New York: Routledge & Kegan Paul.

Sutton, J. (2008). Between individual and collective memory: Interaction, coordination, distribution. *Social Research*, *75*(1), 23–48.

Sutton, J., Harris, C. B., Keil, P. G., & Barnier, A. J. (2010). The psychology of memory, extended cognition, and socially distributed remembering. *Phenomenology and the Cognitive Sciences*, *9*(4), 521–560. https://doi.org/10.1007/s11097-010-9182-y.

Theiner, G. (2013a). Onwards and upwards with the extended mind: From individual to collective epistemic action. In L. Caporael, J. Griesemer, & W. Wimsatt (Eds.), *Developing scaffolds* (pp. 191–208). Cambridge, MA: MIT Press.

Theiner, G. (2013b). Transactive memory systems: A mechanistic analysis of emergent group memory. *Review of Philosophy and Psychology*, *4*(1), 65–89. https://doi.org/10.1007/s13164-012-0128-x.

Theiner, G. (2015). Group-sized distributed cognitive systems. In *The Routledge handbook of collective intentionality*. New York: Routledge.

Theiner, G., & O'Connor, T. (2010). The emergence of group cognition. In A. Corradini & T. O'Connor (Eds.), *Emergence in science and philosophy*. New York: Routledge.

Theiner, G., Allen, C., & Goldstone, R. L. (2010). Recognizing group cognition. *Cognitive Systems Research*, *11*(4), 378–395. https://doi.org/10.1016/j.cogsys.2010.07.002.

Tollefsen, D. P. (2006). From extended mind to collective mind. *Cognitive Systems Research*, *7*(2–3), 140–150. https://doi.org/10.1016/j.cogsys.2006.01.001.

Tollefsen, D., & Dale, R. (2012). Naturalizing joint action: A process-based approach. *Philosophical Psychology*, *25*(3), 385–407. https://doi.org/10.1080/09515089.2011.579418.

Turnstrøm, K., Katz, Y., Ioannou, C. C., Huepe, C., Lutz, M. J., & Couzin, I. D. (2013). Collective states, multistability and transitional behavior in schooling fish. *PLoS Computational Biology*, *9*(2), e1002915.

Tyler, M., Danilov, Y., & Bach-y-Rita, P. (2003). Closing an open-loop control system: Vestibular substitution through the tongue. *Journal of Integrative Neuroscience*, *2*, 159–164.

Weiskopf, D. (2008). Patrolling the mind's boundaries. *Erkenntnis*, *68*, 265–276.

Wilson, R. A. (2000). The mind beyond itself. In D. Sperber (Ed.), *Metarepresentations: A multidisciplinary perspective* (pp. 31–52). New York: University Press.

Wilson, R. A. (2001). Group-level cognition. *Philosophy of Science*, *68*(3), 262–273.

Wilson, R. A. (2004). *Boundaries of the mind: The individual in the fragile sciences: Cognition*. New York: Cambridge University Press.

Wilson, R. A. (2005). Collective memory, group minds, and the extended mind thesis. *Cognitive Processing*, *6*(4), 227–236. https://doi.org/10.1007/s10339-005-0012-z.

6 Collaborative Action (CoAct) Theory

Socially Constructing Shared Knowledge through Mutual Attunement of Shared Affordances

John Teofil Paul Nosek

CONTENTS

INTRODUCTION

Collaborative Action Theory (CoAct), an affordance-based theory based on ecological psychology, explicates how constructing shared affordances provides an alternative model to constructing shared mental models to achieve effective team action (Nosek, 2011). CoAct provides preciseness in what is needed to develop shared affordances to achieve appropriate action. Computer-based applications are integral to achieving effective team action and user-experience design of these computer-based applications is critical to their success. User-experience design is an example of asynchronous work between designer and user to effect appropriate action at the appropriate time. This chapter reviews CoAct and provides a detailed exemplar of how CoAct can be used to achieve preciseness in user experience design.

BACKGROUND

Nosek (2011) explains CoAct in detail and describes initial efforts to test CoAct. This section reviews key elements of CoAct. As actors move within an environment, they discern available informational structures that afford action possibilities. These action possibilities, affordances, are available for the class of actors who have the same potential to discern informational structures that provide these affordances from the same observation point (Gibson, 1979).

> Cognition is the embodied, embedded and always situated process whereby life forms bound to their respective environments in an essential dialectical relationship thrive "to persist and prevail"[1] within the existential spatio/temporal framework defined by their own corporeal dynamics. . . . A life form and its environment constitute a "closed purposive organization" bound by a relationship of mutual influence.
>
> **(Ferreira, 2019, pp. 1, 2)**

Available Affordances versus Perceived and Conceived Affordances

While an affordance is potentially available for the class of actors with certain capacities, a perceived or conceived affordance is what emerges or surfaces for a given member of the class at a specific moment in time as the member moves through the environment to achieve some goal, i.e., while affordances are available to all actors with similar capacities and can be defined statically, perceived and conceived affordances are a subset of available affordances that emerge dynamically based on what the actor is engaged in at the time and what the goal of the actor is. Affordances can be perceived or conceived. When actors perceive and act, they are not self-aware of the action opportunity of the affordance, whereas, when actors conceive and act, they are self-aware of the action opportunity of the affordance. Most actors perceive and then act. To simplify reading, perceived is used in the text, but conceived is usually possible in those circumstances.

Available Affordances

The critical difference between available affordances and perceived affordances is often overlooked and may be the cause of some confusion. This may stem from researchers who have been introduced to Gibson's work on affordances through Norman and Draper's (1986) work in design of human-computer interaction. Affordances, as used in human interface design, are usually inferred to be available to all humans and not dependent on individual differences. Norman is associated with the idea that interfaces should be "user friendly," i.e., so clear that they intuitively afford any human user what to do with the interface at any time.

However, affordances only make sense when actor capacities are coupled with available informational structures within an environment, i.e., actor and environment are inexorably linked, and affordances exist at the intersection of actor capacities and the environment (Gibson, 1979) (see Figure 6.1).

For a given environment, a change in an actor's capacities can change available action possibilities within the environment (affordances); and for a given actor's capacities, a change in the informational structures in the environment changes affordances (see Figure 6.2a and b).

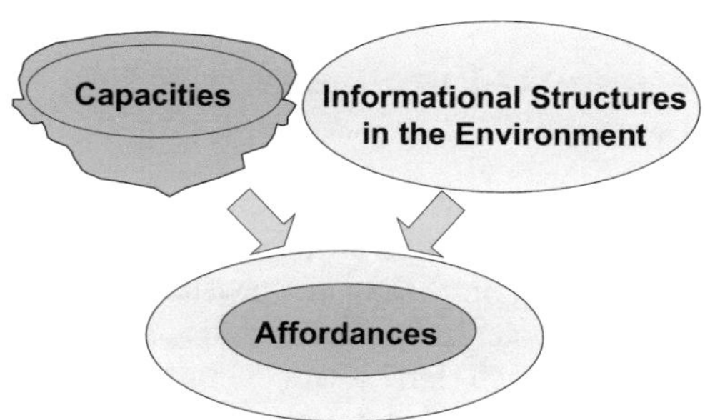

FIGURE 6.1 Affordances = intersection of actor capacities and the environment.

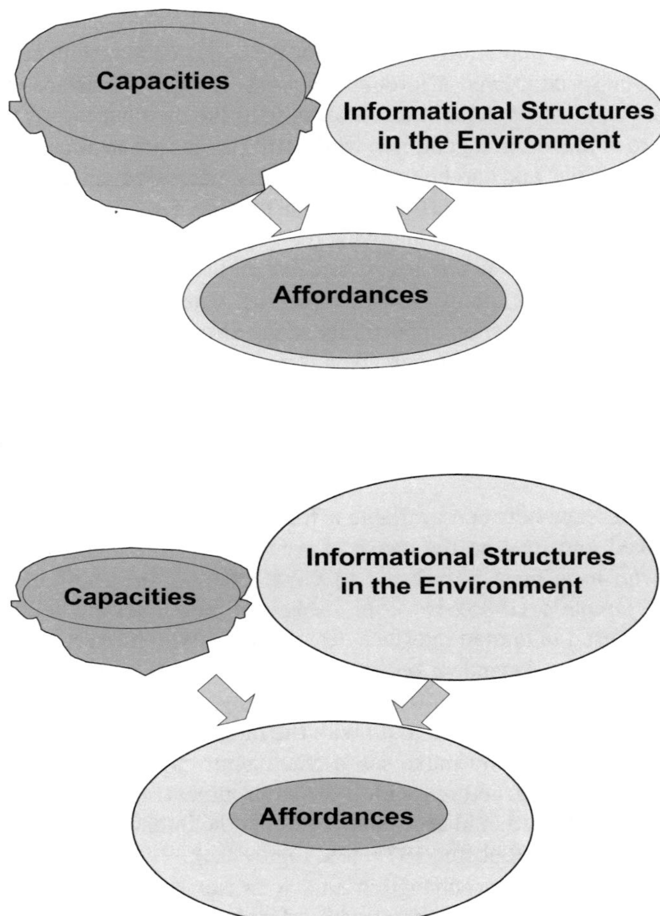

FIGURE 6.2 (a) Changed capacities (increased); (b) changed environment (increased).

EXTENDING GIBSON'S AFFORDANCE-BASED THEORY BEYOND PHYSICAL AFFORDANCES

While Gibson (1979) focused on actor effectiveness based on affordances that are discernable to a class of actors with certain physical capacities to act, CoAct extends the notion of capacities to include more than just physical attributes. This is consistent with Gibson's view of the potentially broad applicability of his theory and the principles of ecological psychology (Heft, 2001). "Gibson's idea of the environment as shared by all perceivers is understood not just for the physical but also for the social environment" (Szokolszky & Read, 2018, p. 20). "An ecological approach presumes a symbiotic relationship between the environment and an agent who is actively seeking out and realizing novel functional relations and therefore is constantly changing the functional environment" (Szokolszky & Read, 2018, p. 9). Szokolsky and Read further argue that Gibson's work should be at the heart of a new, broad science of developmental ecological psychology with the goal to explain "how the organism develops by keeping continuous, active, and reliable contact with its richly structured environment via direct perception and action, over time" (2018, p. 27).

Capacities to act can broadly include capacities to do, think, feel, etc. (Szokolszky & Read, 2018). For example, informational structures in the environment may afford sadness or happiness for one class of actors with certain capacities, but not another. In addition, the idea of an actor moving through an environment is not restricted to the physical movement of the actor within a physical environment but includes the notion that actors are active with respect to what informational structures they pick up, i.e., the notion of static is the exception. Moving through the environment can mean the working through a problem, the reading of a book, the engagement in a discussion. The importance of the concept of moving through the environment means that one is continually experiencing one's environment and is forever changed by this experience, i.e., "cognitively rewired" (Pizlo, 2007). Even visual perception to recognize objects in the environment appears to be dependent on prior experience and/or genetic encoding (Pizlo, 2001, 2007). "Cognitive systems do not passively react to events; they rather actively look for information and their actions are determined by purposes and intentions as well as externally available information and events" (Hollnagel & Woods, 2005, p. 16).

Figure 6.3 depicts how perceived affordances are dynamic and change over time given the goal and current activity. For simplicity, in Figure 6.3 the informational structures within the environment and actor capacities remain the same through time, which means that the set of affordances remain the same for actors in the class with these capacities, while what is perceived within this set changes over time.

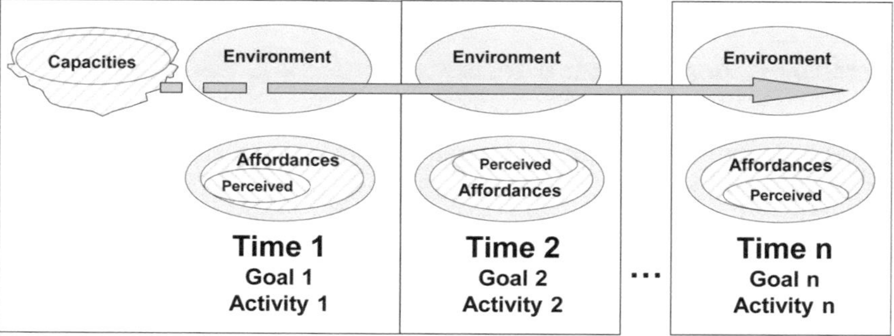

FIGURE 6.3 Perceived affordances = subset of affordances based on goal and current activity over time.

Example of Available versus Perceived Affordances

As noted previously, affordance are action opportunities in the world identified at the intersection of the informational structures in the environment and the capacities of actors. They are static for a given environment and actor capacities and always exist whether perceived or not. Perceived and conceived affordances are those subsets of affordances that emerge when an actor desires to achieve some goal and is engaged in some current activity.

Assume you are sitting in a room. There is a door with a typical doorknob that you used to enter the room. For the typical person of average human height and two arms and hands, in normal conditions, the door affords safe egressibility; the doorknob affords graspability (it would not afford graspability for a one-foot tall person with no arms) and turnability (clockwise or counter-clockwise), and the door affords pushability and/or pullability depending on which way it opens. In normal conditions these affordances always exist for you whether you perceive or conceive them.

When you want to exit the room (goal) and move in the direction of the door (current activity), you grasp the doorknob, turn it, push or pull, and exit. For most of these actions you will not be self-aware of the affordances. For an unfamiliar door you may consciously decide if the door should be pushed or pulled, i.e., conceive of the affordance of pushablity or pullability. Otherwise, these affordances are perceived (you are not self-aware) and you act out these action opportunities.

Assume the door no longer affords safe egressiblity because there is a fire or some danger on the other side of the door. The affordances of doorknob graspability and turnability and the affordance of door pushability or pullability continue to exist for you. However, now you seek safe egressibility and move towards the window. You will not perceive or conceive of doorknob graspability and turnability nor door pushability or pullability. The window affords safe egressibility if you can safely egress through the window, if you can't, then it does not afford safe egressability. As you move to the window, you are more likely conceiving (consciously evaluating) the openability or breakability of the window because of the unfamiliar situation.

You may overestimate your capacities and injure or die egressing the window. In that case the window did not afford you safe egressibility and does not afford safe egressibility for someone with similar capacities.

Collaborative Action (CoAct) Theory

CoAct can be used at multiple levels of granularity, from fine granularity of understanding a single interaction, to tracking intermediate progress and final results of interactions (see Figure 6.4).

For effective collaborative action, actors, whether human or non-human, must perceive shared, relevant affordances at the appropriate time. A shared affordance is an affordance that is shared by more than one actor. To achieve a shared, perceived affordance: (1) the actors must share sufficient capacities so that a given environment affords the same action opportunity for the actors; and (2) this affordance or action opportunity is available (or perceived) at the appropriate time, i.e., an affordance that is available to an actor, but is not perceived when it is supposed to be, is

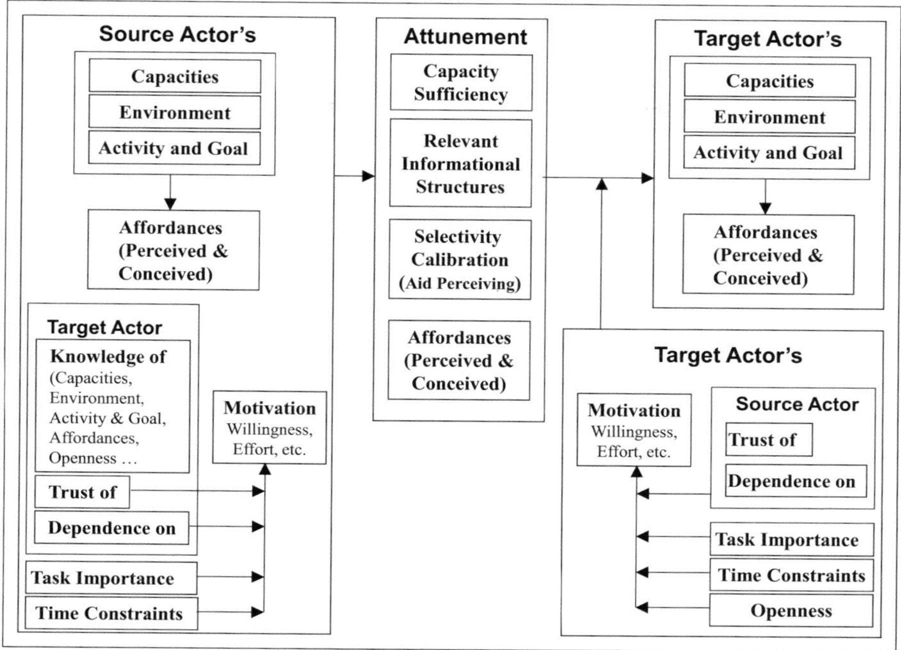

FIGURE 6.4 Model of Collaborative Action (CoAct).

not considered a shared, perceived affordance. A shared, perceived affordance may occur at the same time among actors, but it may also occur asynchronously, as long as this is considered an appropriate time to support collaborative action.

The process to achieve and the end-state of achieving shared, perceived affordances use the same label, "attunement." Attunement (n.d.) means "being or bringing into harmony; a feeling of being 'at one' with another." For example, one can say that collaborators attune each other, i.e., undergo mutually attunement (bringing into harmony) to achieve attunement (being in harmony) of perceiving shared relevant affordances at the appropriate time. Context determines which form of the word is meant.

Actors engaged in collaborative action must take responsibility for mutual attunement by (1) sharing relevant informational structures, (2) bringing each other up to the sufficient capacities of the class of actors for whom the environment affords the desired, relevant action opportunities (affordances), and (3) assisting each other in perceiving the subset of relevant affordances at the appropriate time (selectivity calibration).

In Figure 6.5, we identify the actor who is actively attuning as the Source Actor, and the actor who is the target of this attunement effort as the Target Actor.

However, these roles alternate as collaborators engage in mutual attunement. Initially, at Time 1, the actors do not share the relevant informational structures

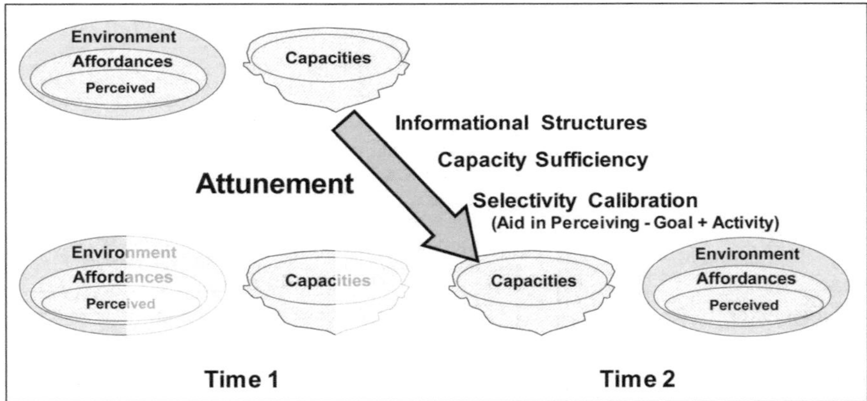

FIGURE 6.5 Attunement to perceive shared, relevant affordances at the right time.

within the environment and sufficient capacities. Therefore, they cannot perceive relevant affordances. This is depicted by the actor at the bottom (the Target Actor) only sharing half of what the actor on the top (Source Actor) has available.

In Time 2, the Source Actor attunes the Target Actor by sharing relevant informational structures in the environment, assisting the Target Actor to achieve sufficient capacities (this is shown by the capacities of the Target Actor are now the same as the Source Actor), and assisting the Target Actor in perceiving relevant informational structures at the appropriate time (selectivity calibration). At the end of Time 2, both actors are perceiving similar, relevant affordances. For simplicity's sake, although we only depict one Target Actor in Figure 6.5, but there could be more than one Target Actor in a given collaborative act.

In Gibsonian terms, actors attune each other to perceive relevant affordances by building sufficient capacities in each other, exchanging informational structures so they are available to both, and bringing each other to a common observation point within the environment of available informational structures at the appropriate time (see Figure 6.6).

Attunement can be intended and unintended, explicit and implied, verbal and nonverbal. For example, an actor may explicitly expose his/her identity intending to imply positive informational structures as to status, however, the receiving actor may not treat this datum as positive. As noted earlier, actors attune each other by sharing relevant informational structures in the environment, assisting each other to achieve sufficient capacities, and assisting each other in perceiving relevant informational structures at the appropriate time. Table 6.1 summarizes components of attunement.

Note, not all components exist in every act of attunement. Attunement related to selectivity calibration aligns the goals and activities of actors. This relates to Gibson's idea that an actor moves within the environment picking up informational structures to achieve some goal and affects what affordances are perceived at a given moment in time.

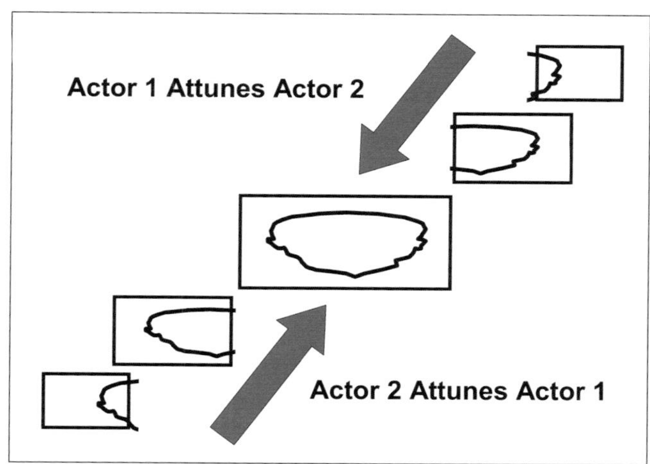

FIGURE 6.6 Gibsonian mutual attunement.

TABLE 6.1
Summary of Attunement Components

Informational Structures: Relevant aspects of the environment/situation.

Capacity Sufficiency Building: Building sufficient, similar capacity needed so that affordances needed for collaboration are available to the actors.

Perceived Affordances: Perceived action opportunities in the environment (inferred through observation).

Conceived Affordances: Conceived action opportunities in the environment (explicit through self-aware reflection).

Selectivity Calibration: Assistance in perceiving the relevant affordances at the appropriate time.

Achieve goal or fulfill need—align goals; **Current activity**—align current activity.

One aspect of the model may not be adequately depicted but deserves special mention. Target Actors may accept the affordances of the Source Actor without being brought up to sufficient capacities and having shared informational structures so that these affordances are available for Target Actors themselves. For example, Source Actors, who are trusted, may be able to transfer their affordance, but the Target Actors do not have sufficient capacities so that they can perceive the affordances for themselves.

If Source Actors expose their affordances (perceived and conceived) as part of their attunement for a given shared environment, is this a form of learning? That is, does this teach the Target Actor, either explicitly through conceived affordances transmitted or implicitly through perceived affordances ascertained through observation, that these and future, similar informational structures afford these action possibilities? Does repetition reinforce building sufficient capacities? Is learning by example a form of this process?

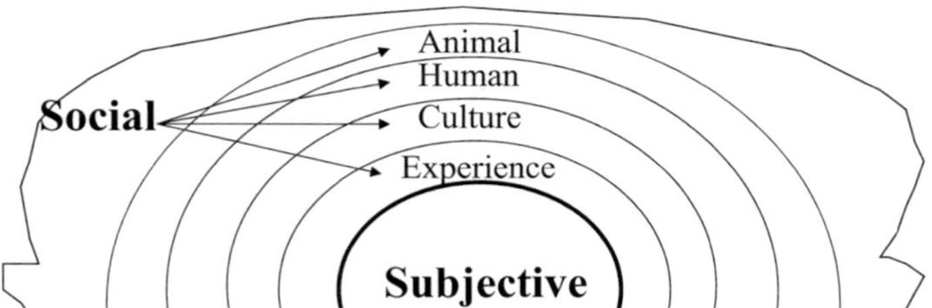

FIGURE 6.7 Example of stratifying the shared social world of an individual actor.

Note, attunement is neither inherently good nor correct and can result in building sufficient capacities so that Target Actors perceive good, correct affordances or warped, incorrect affordances (Boyle, Kacmar, & George, 2008). For example, Source Actors who enjoy very high trust, such as some fanatical religious and political leaders, may be able to transfer their affordances of world events in a continuously, warped manner such that Target Actors will perceive affordances within current and future environments that will result in unwarranted and apparently irrational actions, but consistent with CoAct.

Habermas (1984) offers the concept of lifeworlds to explain the difficulty in achieving shared affordances among actors from different lifeworlds. Borrowing loosely from Habermas, actors of dissimilar backgrounds may be able to find some common elements of lifeworlds that can be shared. Figure 6.7 provides an example of stratifying the shared social world of an individual actor.

The unshared subjective world can only be contained within the shared social world. An actor living in the world is part of and helps to create a shared social world.

For example, the Target Actor may be an integral part of the decision making process but is not experiencing the situation for himself or herself and may not have the ability to understand the relevancy of facts of the situation at hand. In this case, actors must find a common aspect of a lifeworld to share, for example, the capacities of actors from the two different classes may intersect on similar cultural or human capacities to pick up informational structures to share that the situation affords danger. Two examples are offered to illustrate this.

Example 1: Command and Control

In a military situation, a commander on the ground feels threatened, perceives the attackability of the enemy, but must obtain permission to engage from a distant political leader with no military background. In this case, the Source Actor from the class of experienced warfighters must obtain permission from this politician, the Target Actor, who is not within the class of experienced warfighters, and may not be attunable within limited time requirements, to perceive the affordance of attackability of the enemy. However, the Source Actor (warfighter) can then expose behavior that will permit the political decision maker to perceive that the situation affords danger. The decision maker at headquarters would then trust the local on-scene commander to make the best decision (including attacking) and implicitly trust the advancement system (shared culture) that puts commanders in this position of responsibility.

Example 2: Medical Emergency Surgery

In emergency exploratory surgery, family members and patients may share decision-making responsibility with surgeons but do not have the capacities of the surgeon to pick up informational structures that afford necessary surgery. It may be even more difficult for family members than for patients, who may be experiencing pain, to perceive affordances available to the class of surgeons. Surgeons should, and the best most likely do, focus on selectivity calibration and informational structure sharing to achieve the shared affordance of imminent danger among surgeon, patient, and family members. Patients and family members must then trust the surgeon to make the best decision and implicitly trust the medical system (shared culture) that puts surgeons in this position of responsibility.

Example Applications of CoAct

In addition to the examples in group decision making provided above, two examples of other applications of CoAct include user experience design and information overload/reduced bandwidth.

User Experience Design

Humans, using computers to complete their work, progress through a sequence of cycles of interaction steps and information displays, where the last information display provides affordances for the next interaction step as the user acts to achieve some goal. For each step in the cycle, the designer assumes an envisioned user, i.e., the design will provide appropriate affordances at the appropriate time for a user with certain capacities engaged in an activity to achieve some goal. Differences between the envisioned user and the actual user determines design quality and attunement needed, such as training for the actual user.

Reducing Data Overload/More Effective Use of Bandwidth

Because some data may provide more informational value that other data, in times of reduced bandwidth, CoAct may provide guidance on prioritizing what data should be transmitted or received.

USING COACT TO PRECISELY DESIGN USER EXPERIENCE—EXEMPLAR

In this exemplar, CoAct and Claims Analyses (Rosson & Carroll, 2002) are integrated to assess usability of a computer-based interface, to make informed recommendations for improving usability, and to evaluate the usability of a redesigned interface. This process allows developers and usability experts to uncover key usability issues in their design and to determine if the interface satisfies usability requirements for their audience. The evaluation also helps to locate opportunities for just-in-time training to mitigate the need for the user to hold specific semantic and syntactic knowledge prior to using the system.

This exemplar is divided into the following sections: (1) an introduction to the utilized tools and methods, (2) an initial evaluation of a simple interface feature and redesign recommendations, (3) an evaluation of the redesigned feature, and (4) a usability comparison between the original and redesigned features.

Tools and Methods

CoAct Analysis

CoAct Analysis provides a technique to assess the usability of interface features involved in completing a given activity. This involves identifying an activity to analyze and then proceeding through the activity until completion while determining the affordances provided by the interface at each step. An expanded version of the CoAct Analysis which also evaluates effectiveness of the design in providing certain affordances was used. Figure 6.8 presents the template that was used in this analysis.

The following list briefly describes the fields within the template:

- **Activity**: A specific activity supported by the interface. Usability of the interface in achieving completion of this activity will be assessed.
- **Emerging Goals/Subgoals:** The smaller tasks involved in completing the larger activity. Consideration of the following questions for each subgoal:
  - *Is the first step to take to achieve the subgoal obvious?*
  - *After achieving the subgoal, is the next step after that obvious?*
- **Affordances:** The possible actions that are available to the user based on the capacities of the user and the design of the interface, both of which we consider in further detail:
  - *Capacities of the User*
    - **Semantic Knowledge:** The conceptual knowledge needed by the user to make an affordance available.
    - **Syntactic Knowledge:** The knowledge of specific steps needed to be completed by the user to make an affordance available.
  - *Design of the Interface*
    - **Design to Achieve**: The interface design and features that allow an available affordance to be recognized by the user.
    - **Goodness of Design**: The effectiveness of the design to achieve and any possible improvements that may allow an affordance to be more readily recognized by the user.

Claims Analysis

The Claims Analysis provides a methodology for assessing specific design features of an interface and determining if their design effectively supports their intended function. This analysis deals specifically with the organization and aesthetics of the features and relies on Gestalt principles to determine which interface features effectively support usability of the interface. This analysis can provide insight for both the current design and suggested redesigns, and considers the implementation involved in a redesign. Figure 6.9 presents the template used in this analysis.

CoAct Analysis	*Activity title*					*Page Title*					
Q1: Is the first step to take to achieve the subgoal obvious?											
Q2: After achieving the subgoal, is the next step after that obvious?											
Screen #	**Informal Cognitive Walkthrough**		**Emerging Goal/Subgoal**	**Affordance**	**Essential Capabilities (Knowledge) Required**						
	Q1	*Q2*			*Semantic*	*Design to Achieve*	*Goodness of Design*	*Syntactic*	*Design to Achieve*	*Goodness of Design*
1											
2											
3											

FIGURE 6.8 CoAct Analysis Template.

Claims Analysis	*Activity title*		*Page Title*		
Feature #	**Design Feature**	**Claims**		**How to Impement**	**Difficulty to Impement**
		Positive	*Negative*		
1					
2					
3					

FIGURE 6.9 Claims Analysis Template.

The following list briefly describes the fields within the template:

- **Design Feature:** A specific feature included in the interface.
- **Options:** A specific design option for the feature, either the current design or a recommended redesign.
- **Claims**: Statements about the design of an interface feature, including:
  - **Positive Claims:** Those that demonstrate a design supports usability.
  - **Negative Claims:** Those that demonstrate a design does not support or detracts from usability.
- *For possible redesign options, the analysis considers the following from the development perspective:*
  - **How to Implement:** A brief description of the development needed to implement the redesign.
  - **Difficulty to Implement:** A rough estimate of the difficulty of the development needed to implement the redesign (e.g., low difficulty, moderate difficulty).

Analysis Results of Specific Feature

The Feature

For this exemplar, a simple interface feature of the Virtual Teaching Assistant (VTA) web application was analyzed. VTA provides an interface for autism therapy providers to perform therapy management and progress tracking of their students. On the Therapy Assignment Page, VTA users can perform the activity of assigning and/or updating therapy programs and behaviors for a student. Selecting a student emerges as a subgoal involved in completing this activity. The analysis that follows will focus specifically on the interface feature that allows a user to select a student during the therapy assignment activity.

Original Design

The original design feature for student selection combines an input box with a dropdown list (see Figure 6.10). The input box allows a user to search for a specific student, while the dropdown list allows the user to view all available students.

CoAct Analysis—Original Design

The following sections highlight key information gathered from the CoAct Analysis, including affordances provided by the interface, semantic and syntactic knowledge needed by the user to utilize these affordances, and the effectiveness of the interface design in supporting user recognition of these affordances.

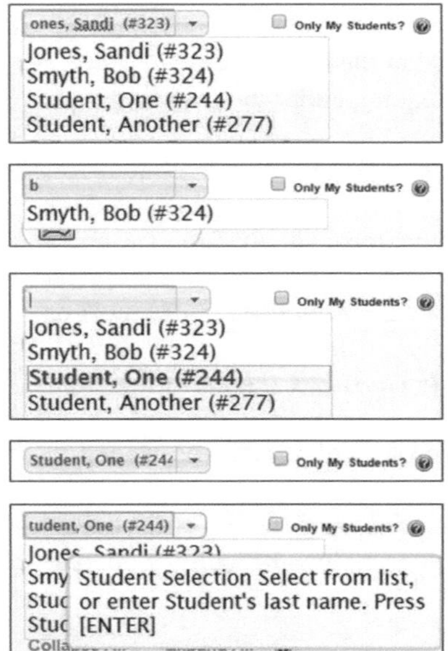

- Clicking the dropdown arrow opens a dropdown list of all available students.
- Typing in the input box opens a dropdown list of filtered results.
- Hovering over a student name with the cursor highlights the active student name and makes it available for selection.
- Clicking the name selects that student.
- The tooltip overlaps the dropdown list of students.

FIGURE 6.10 Original design.

Affordances—Original Design

The following list highlights the affordances provided by the interface for the user with the necessary semantic and syntactic knowledge:

- Revealability of the tooltip for selecting and searching students
- Revealability of available students
- Searchability of student's name
- Selectability of student
- Clearability of previously selected student's name

Semantic and Syntactic Knowledge—Original Design

In general, most of the semantic and syntactic knowledge needed to make these affordances available is possessed by the average computer user, including understanding the conceptual processes and executing the steps involved in moving the cursor over a feature and interacting with the feature through clicking. However, analysis reviews a few key areas in which semantic and syntactic knowledge needed by the user could be reduced or simplified:

- Upon first use of the Therapy Assignment Page, the user must know that the input box can be used to search for students. This knowledge can be acquired through the tooltip, yet opening the tooltip requires even further semantic and syntactic knowledge.

- Upon continued use of the therapy assignment page, the user must know that the student's name in the input box can be removed by backspacing the text and that another student's name can be entered. This knowledge cannot be readily acquired or understood through the interface design.
- The user must know that a tooltip with further instructions will open upon hovering over the dropdown arrow. The tooltip's connection to the arrow may be beneficial when a user decides to use the dropdown arrow, and then receives further information. However, if a user does not attempt to use the dropdown arrow, the knowledge of the tooltip or the knowledge provided by the tooltip will not be readily available.
- Since the tooltip contains information for both methods (input box search or dropdown list selection) of selecting a student, revealing the tooltip should require the same semantic and syntactic knowledge when using either method to simplify the knowledge needed to utilize the tooltip.
- Since using either method (input box search or dropdown list selection) leads to completion of the same goal (selecting a student), revealing the list of available students should require the same semantic and syntactic knowledge when using either method. In general, the methods for selecting a student should be more tightly integrated to simplify the knowledge needed to select students.

Goodness of Design—Original Design

The above issues related to semantic and syntactic knowledge needed by the user can be remedied thorough improving the design of the interface, especially in providing the user more thorough just-in-time training. The following list highlights issues related to goodness of design:

- Upon first use of the therapy assignment page, the input box should not be blank. Rather, the blank space should be used to provide brief just-in-time training for the user.
- Upon continued use of the therapy assignment page, the interface should provide clear visual indication that the name of a previously searched student can be removed to perform a new search. The removal process should require fewer steps than current design of manual backspacing.
- An expanded area should be able to trigger the tooltip. This allows the just-in-time training that it provides to be more easily accessible, no longer requiring the user to navigate to a highly specific part of the screen (i.e., the dropdown arrow).
- The tooltip should be positioned such that it no longer overlaps the dropdown list of students. Since it currently interferes with functionality, a user may find it disruptive and be less inclined to benefit from the just-in-time training that it provides.
- The tooltip offers the most thorough just-in-time training for the user. Therefore, its instructions should be more comprehensive to benefit all levels of users.

- The combination box should be more tightly integrated so the process of selecting a student becomes more specific and streamlined. Currently, the combination box seems to offer two functions: student searching using the input box and student selection using the dropdown list. Rather, the interface should approach this feature to perform one coherent action: selecting a student.

Claims Analysis—Original Design

The following sections highlight important positive and negative claims made for both the original interface design, as well as for possible redesigns which aim to solve specific issues identified in the CoAct Analysis. The analysis also considers implementation factors for each redesign option.

Claims—Original Design

Performing a Claims Analysis on both the original design and possible redesigns allows for the most effective interface features to be chosen based on the ratio between positive and negative claims. For most features, the analysis allowed for identification of designs that eliminated negative claims. The following list describes the claims made for each feature in the interface for both the original design and any effective redesigns:

- *Tooltip for student selection/tooltip for student searching*
  - **Current Design**: Tooltip triggered when user hovers on down arrow
    - **Positive Claims:** Does not clutter screen.
    - **Negative Claims**: Just-in-time training not initially visible, user must hover over specific area to trigger tooltip, tooltip may be overlooked by a user that does not use the down arrow, tooltip overlaps dropdown list, and limited instructions provided by the tooltip.
  - **Effective Redesign:** Tooltip triggered when user hovers on entire combination box, with a placeholder in the blank input box, repositioning of the open tooltip, and revised tooltip content.
    - **Positive Claims**: Does not clutter screen, placeholder provides brief just-in-time training, tooltip will be triggered when the user interacts with any area of the combination box, tooltip does not interfere with the dropdown list, and comprehensive instructions provided by the tooltip.
    - **Negative Claims**: None.
- *Student selection dropdown/ student search box*
  - **Current Design**: Dropdown list opens when user clicks down arrow or user begins entering text to search for student
    - **Positive Claims**: Down arrow often associated with dropdown lists and input boxes often associated with text input.
    - **Negative Claims**: Must either begin searching or click a specific area (i.e., down arrow) to see all available students, the dropdown list and search functions not seamlessly integrated, and no visual indication that a previous student name can be removed to begin a new search.

- **Effective Redesign**: Dropdown list opens when user clicks input box, with down arrow removed from the design and an (x) button in the input box to easily clear the previous search.
    - **Positive Claims**: Provides only one simple course of action for interacting with the input box (i.e., clicking into it) which automatically opens list of available students, tightly integrates dropdown list and searching into one function, and provides a visual indication that a previous student name can be removed to begin a new search.
    - **Negative Claims**: None.

IMPLEMENTATION REQUIREMENTS

All redesigns can be implemented through simple changes in HTML, CSS, and JavaScript. Implementing the (x) button for clearing a search term from the input box may be slightly more involved. However, existing implementation solutions are readily available from open source websites. All suggested redesigns could be classified as low difficulty implementation. Thus, implementation factors did not affect the redesign recommendations.

Recommendations

Based on findings from the CoAct and Claims Analyses, the following redesigns are recommended to increase usability of the interface:

- Add a placeholder in the empty input box that reads "Select or search a student."
- Remove down arrow, leaving only the input box.
- Trigger tooltip when user hovers over input box.
- Add more comprehensive instructions to the tooltip: "Click to see list of available students. Enter student's name to search. Click student name to select. Click (x) to clear student name and begin new search."
- Reposition tooltip such that it does not overlap the dropdown list.
- Add (x) button to input box to clear previously selected student.

Redesign

The redesign incorporates recommendations derived from the CoAct and Claims Analyses performed on the original design (see Figure 6.11). The redesigned feature provides an input box that allows the user to view all available students from a dropdown list as well as search for a specific student.

CoAct Analysis of Redesign

The following sections highlight key information gathered from the CoAct Analysis of the redesign, including affordances provided by the interface, semantic and

- Add a placeholder in the empty input box that reads: "Select a student."
- Remove the down arrow, leaving only the input box.
- Trigger the tooltip when a user hovers over input box.
- Add more comprehensive instructions to the tooltip.
- Reposition the tooltip such that it does not overlap the dropdown list.
- Open the dropdown list when user clicks input box.
- Add an (x) button to the input box to clear previously selected student.

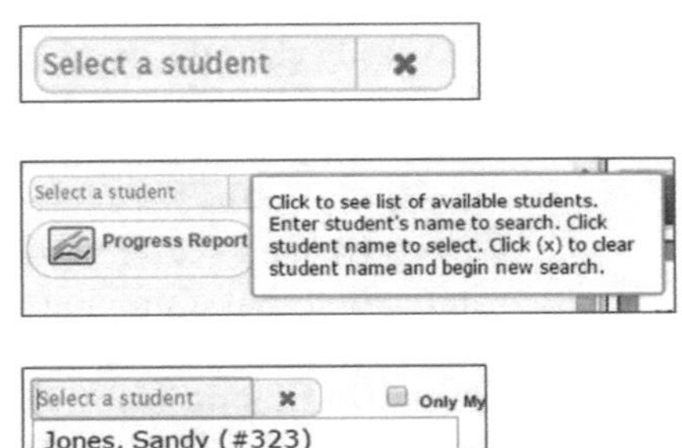

FIGURE 6.11 Recommended redesign.

syntactic knowledge needed by the user to utilize these affordances, and the effectiveness of the interface design in supporting user recognition of these affordances.

Affordances—Redesign

The following list highlights the affordances provided by the redesigned interface:

- Revealability of the tooltip for selecting and searching students
- Revealability of available students
- Searchability of student's name
- Selectability of student
- Clearability of previously selected student's name

Semantic and Syntactic Knowledge—Redesign

When you compare the semantic and syntactic knowledge required in the original design to the redesign, to make similar affordances available the knowledge requirement has been reduced. The following list describes the semantic and syntactic knowledge needed by the user to make affordances available in the redesigned interface:

- Interacting with the input box will allow for student selection.
- Clicking the input box will open a list of available students.
- Typing in the input box will filter the list of available students.
- Moving the cursor over a student's name and clicking will select that student.

- Clicking the (x) button will clear the previously selected student in order to search again.

Goodness of Redesign

The semantic and syntactic knowledge requirements of the user have been optimized through the good design of the interface. The design reduces prior semantic and syntactic knowledge required of a user and therefore *expands the potential number of users with sufficient existing semantic and syntactic knowledge to complete their work*. The design reduces the need for a priori training. Incorporating just-in-time training decreases the knowledge needed in advance by a user to utilize the interface, facilitates user perception of the affordances available in the interface, and ultimately guides the user through an activity. The following list highlights the design features that contribute to goodness of design:

- Upon first use of the therapy assignment page, the input box contains a placeholder that reads "Select or search a student." The placeholder prompts the user to interact with the input box in order to search or select a student. The user must only have semantic and syntactic knowledge related to moving the cursor.
- Upon interacting with the input box after the initial prompt, a tooltip with comprehensive instructions immediately opens and provides the following training for the user: "Click to see list of available students. Enter student's name to search. Click student name to select. Click (x) to clear student name and begin new search." Thus, when a user chooses to interact with the tool for student selection, the interface offers the semantic and syntactic knowledge needed to perform any action related to student selection: viewing the available students, searching a student, selecting a student, and beginning a new search. Again, the user must only hold knowledge related to moving the cursor, clicking, and typing with the keyboard.
- In addition to the verbal instruction in the tooltip, the (x) button provides clear visual indication that the name of a previously searched user can be removed. This feature simplifies the knowledge needed by the user to perform subsequent student searches by reducing the action of clearing a previously selected student into one step (i.e., clicking the (x) button).

Claims Analysis—Redesign

The following section highlights important claims made for the redesigned interface. The Claims Analysis also considers Gestalt principles that "describe the configural properties of visual information" (Rosson & Carroll, 113). When used successfully in an interface, "the principles of Gestalt perception direct attention to the many relationships among information elements in a complex display" (Rosson & Carroll, 114). Therefore, effective use of Gestalt principles in interface organization contributes to positive claims made about an interface.

Claims—Redesign

The redesign resolves issues present in the original design, thus reducing negative claims for the student selection feature. The Claims Analysis for the redesign focuses on positive claims and demonstrates the redesign's consistency with Gestalt principles. The following list describes the positive claims made and Gestalt principles followed for each feature in the interface:

- *Tooltip for student selection/tooltip for student searching*
  - **Current Design:** Tooltip triggered when user hovers input box, with a placeholder in the blank input box, repositioning of the open tooltip, and revised tooltip content.
    - **Positive Claims**: Does not clutter screen, placeholder provides brief just-in-time training, tooltip will be triggered when the user interacts with any area of the input box, tooltip does not interfere with the dropdown list, and just-in-time training provided by comprehensive instructions in the tooltip.
    - **Gestalt Principles**: *Proximity* (tooltip opens near the input box), *Closure* (placeholder located within the input box, encapsulating the just-in-time training within the figure), and *Area* (placeholder located within the input box, reducing the size of the overall figure). These principles indicate the relationship between various parts of the figure.
- *Student selection dropdown/ student search box*
  - **Current Design**: Dropdown list opens when user clicks input box and (x) button in the input box clears the previous search.
    - **Positive Claims**: Provides only one simple course of action for interacting with the input box (i.e., clicking into it) which automatically opens a list of available students, tightly integrates using dropdown selection and searching into one function, and provides a visual indication that a previous student name can be removed to begin a new search.
    - **Gestalt Principles**: *Closure* (the input box, the (x) button, and the dropdown list all form one closed figure), *Area* (the (x) button resides within the input box, and the dropdown list opens immediately beneath the input box, reducing the space used by the overall figure), and *Similarity* (student names in the dropdown list are formatted consistently). These principles indicate the relationship between various parts of the figure and reveal when certain elements (i.e., student names) represent the same information type.

Discussion of Original Design and Redesign Using CoAct

The following discussion compares findings from the analyses performed for both the original design and the redesign and explores the effect of the analyses on improving key elements of the interface. The section addresses affordances, semantic and syntactic knowledge needed by the user to utilize each interface, and just-in-time training offered by each interface to increase usability.

Have the Affordances Changed?

Based on the CoAct Analyses, the affordances remain the same for both the original design and the redesign:

- Revealability of the tooltip for selecting and searching students
- Revealability of available students
- Searchability of student's name
- Selectability of student
- Clearability of previously selected student's name

Has Semantic and Syntactic Knowledge Been Reduced?

Yes. In the original design, the user needed more extensive semantic and syntactic knowledge for the affordances to be available. For example, the user needed to initially understand without any prompt that:

- The input box provided for student searching.
- The dropdown arrow triggered a tooltip with further instructions.
- Backspacing the name of a previously selected student would allow for another search.

The user should not be required to hold this knowledge before using the interface. Instead, the interface should reduce the need for previous knowledge to a minimum through just-in-time training. The redesign effectively uses just-in-time training to decrease the knowledge required by the user to do the following:

- Move the cursor
- Click a feature
- Use the keyboard to type

Thus, the redesign allows for users with a minimum knowledge to utilize the interface efficiently.

Has Just-in-Time Training Been Improved?

Yes. In the original design, the interface did provide just-in-time training within the tooltip. However, revealing the tooltip required the prior knowledge that hovering over the dropdown arrow would trigger the tooltip. If the user did not interact with the dropdown arrow, the just-in-time training would not be readily accessible. In addition, the tooltip did not contain entirely clear instructions.

To improve the effectiveness of just-in-time training, a minimum amount of knowledge should allow the user to access it for all levels of users to benefit from the knowledge increase that it provides. Therefore, the user should only be required to hold basic semantic and syntactic knowledge (e.g., moving the cursor, clicking) with just-in-time training and visual cues bridging the gap between the user's knowledge and the required knowledge to utilize the interface. The following interface features of the redesign support just-in-time training:

- The placeholder in the input box instructs the user that the student selection feature can be used to select or search students and acts as a visual cue to interact with this feature if the user aims to perform the action of student selection. Without this brief training, the user might not hold the knowledge that this feature performs the specified activity.
- The placeholder acts as a prompt to interact with the student selection feature. When the user interacts with any area of the input box, the interaction triggers a tooltip with further instructions to open. Thus, when the user demonstrates an interest in using this feature, the interface immediately offers the user the knowledge required to perform any action afforded by the feature: viewing the available students, searching a student, selecting a student, and beginning a new search.

The just-in-time training provided by the redesigned interface reduces the a priori semantic and syntactic knowledge needed by the user to a minimum and delivers further instructions when the user needs to perform tasks requiring greater than minimum knowledge. Thus, the training equips all users with the knowledge required to use this interface feature, regardless of prior experience using the application.

SUMMARY

CoAct extends the notions of affordance and moves away from idiosyncratic, subjective mental models of the world to the notion that actors with similar capacities to act can potentially discern similar action possibilities in the world. CoAct has the potential to impact user experience design, provide guidance on reducing data overload among classes of users with different capacities to act, and improve collaborative decision making. In this chapter, an exemplar was provided where CoAct Analysis was integrated with Claims Analysis to improve user experience design. CoAct and Claim Analyses of the original design identified precision in what semantic and syntactic knowledge was required to make affordances available and identified areas for design improvement. CoAct and Claims Analyses of the redesign were used to precisely identify the reduced semantic and syntactic knowledge required to make the same affordances available and how redesign will aid in helping users to perceive available affordances.

ACKNOWLEDGMENTS

It was a pleasure to work with Sharon Tartarone in completion of the exemplar. She quickly grasped CoAct and provided a clear analysis of how CoAct can provide precision in user experience design.

REFERENCES

attunement. (n.d.). *Dictionary.com's 21st Century Lexicon*. Retrieved September 25, 2009, from Dictionary.com website http://dictionary.reference.com/browse/attunement,

Boyle, R. J., Kacmar, C. J., & George, J. F. (2008). Distributed deception: An investigation of the effectiveness of deceptive communication in a computer-mediated environment. *International Journal of e-Collaboration*, *4*(3), 14–39.

Ferreira, M. I. A. (2019). Cognitive architectures: The dialectics of agent/environment. In M. I. A. Ferreira, J. S. Sequeira, & R. V. Ventura (Eds.), *Intelligent systems, control and automation: Cognitive architectures* (Vol. 94, pp. 1–12). Switzerland: Springer Nature.

Gibson, J. J. (1979). *The ecological approach to visual perception*. Boston, MA: Houghton-Mifflin.

Habermas, J. (1984). *The theory of communicative action: Reason and the rationalization of society*. Boston, MA: Beacon Press.

Heft, H. (2001). *Ecological psychology in context*. Mahwah, NJ: Laurence Erlbaum.

Hollnagel, E., & Woods, D. D. (2005). *Joint cognitive systems*. New York: CRC Press.

Norman, D. A., & Draper, S. W. (1986). *User centered system design*. Mahwah, NJ: Laurence Erlbaum.

Nosek, J. T. (2011). Towards an Affordance-based Theory of Collaborative Action (CoAct). *International Journal of e-Collaboration*, *7*(4), 37–60.

Pizlo, Z. (2001). Perception viewed as an inverse problem. *Vision Research*, *41*(24), 3145–3161.

Pizlo, Z. (2007). Human perception of 3D shapes. In W. G. Kropatsch, M. Kampel, & A. Hanbury (Eds.), *Computer analysis of images and patterns* (Vol. 4673, pp. 1–12). Berlin: Springer.

Rosson, M. B., & Carroll, J. M. (2002). *Usability engineering: Scenario-based development of human computer interaction*. New York: Morgan Kaufmann.

Szokolszky, A., & Read, C. (2018). Developmental ecological psychology and a coalition of ecological–relational developmental approaches. *Ecological Psychology*, *30*(1), 6–38. https://doi.org/10.1080/10407413.2018.1410409.

7 Mismatches between Perceiving and Actually Sharing Temporal Mental Models

Implications for Distributed Teams

Jacqueline Marhefka, Susan Mohammed, Katherine Hamilton, Rachel Tesler, Vincent Mancuso, and Michael D. McNeese

CONTENTS

Distributed or virtual teams are becoming increasingly prevalent in occupational areas ranging from legal services to software development to hospital and medical services (Gilson, Maynard, Young, Vartiainen, & Hakonen, 2015; Kimble, 2011). Compared to teams with more face-to-face interaction, distributed teams face unique challenges due to the absence of normally salient visual, auditory, and social team member cues that facilitate positive group dynamics (Kimble, 2011). As a result, virtual team interactions may deteriorate into process loss in which the team fails to live up to its potential (Schmidtke & Cummings, 2017). Due to the lack of opportunity for unspoken knowledge sharing through social and physical cues, there is greater difficulty in processing task information in virtual teams (McLeod, 2013; Rentsch, Delise, Mello, & Staniewicz, 2014) and less open information sharing (Bazarova & Walther, 2009; Mesmer-Magnus, DeChurch, Jimenez-Rodriguez, Wildman, & Shuffler, 2011). The lack of face-to-face interactions may contribute to coordination breakdowns and mistrust (McLeod, 2013). Lack of trust, in turn, may contribute to biases in interpretations and judgments about other members, hampering understanding of team member behavior (Bazarova & Walther, 2009).

How members "get on the same page" regarding what will be accomplished and how team duties will be managed is a formidable challenge in any team context, but especially in distributed teams. Distributed cognition examines how cognitive systems are organized both between people and with resources in the virtual environment (Hollan, Hutchins, & Kirsh, 2000). Despite agreement that it is important to investigate, research has given minimal attention to how shared cognition develops in virtual teams (Maynard & Gilson, 2014; Schmidtke & Cummings, 2017). As a specific type of shared cognition, we focus on team mental models (TMMs) or the shared, organized understanding of the key elements in a team's environment (Mohammed, Ferzandi, & Hamilton, 2010). According to Ellwart, Happ, Gurtner, and Rack (2015), it is critical, particularly for virtual teams, to synchronize thinking patterns and expectations to develop a shared TMM. Throughout the chapter, we focus on sharedness or similarity as the primary property of TMMs (Mohammed et al., 2010). Whereas TMMs describe similarity on taskwork, teamwork, or "timework" (Mohammed, Hamilton, Tesler, Mancuso, & McNeese, 2015), shared situational awareness addresses members' shared understanding of the current situation (Wellens, 1993).

In the early 1990s, the concept of TMMs was developed to capture implicit coordination within a team (Cannon-Bowers, Salas, & Converse, 1993; Klimoski & Mohammed, 1994). Members with high TMM similarity are able to anticipate each other's behaviors and needs (Cannon-Bowers et al., 1993). Over two decades of conceptual, measurement, and empirical research has solidified the TMM construct as a unique and impactful form of team cognition (Mohammed et al., 2010). TMMs are recognized as an important contributor of team effectiveness

(Mohammed et al., 2010) due to their impact on team processes such as coordination (Salas, Sims, & Burke, 2005) and outcomes including performance (DeChurch & Mesmer-Magnus, 2010a, 2010b). Despite substantive progress in TMM research, several research needs continue to be identified, including the awareness of sharing among team members (Mohammed et al., 2010).

Specifically, what happens when team members think they are in agreement but are not in actuality? Discrepancies between perceptions and reality can exist in any team, but particularly for distributed teams. Due to the difficulty in sensing team members' contexts and motives (Bazarova & Walther, 2009) as well as the lack of visual and social cues (Rentsch et al., 2014), it is likely that virtual teams will struggle to accurately detect perceived and actual TMM similarity. The misperceptions that result may prove detrimental to performance and continued member interactions. In response to this special challenge in distributed teams, we examined the relationship between perceived and actual TMM similarity on team performance and viability (willingness of team members to work together in the future). We focused specifically on temporal TMMs, common views held by the team about the time-related aspects of performing collective tasks (Mohammed et al., 2015).

Theoretically, this study helps to resolve a long-standing debate in the TMM literature regarding the role of perceptions of similarity (Mohammed et al., 2010). Because most TMM research has examined only actual sharedness, little is known about mismatches between perceived and actual TMM similarity (Rentsch, Small, & Hanges, 2008). Empirically, we simultaneously assess actual and perceived TMM similarity, whereas previous studies have examined either, but not both. Practically, this research could help to diagnose teams that think they are in sync, but are not in actuality, paving the way for targeted and tailored training.

THEORETICAL BACKGROUND

Team Mental Models

TMMs are a specific form of team cognition, a broader concept describing how knowledge is gathered and held within a team. TMM similarity describes how convergent or consistent team members' mental models are with each other (Rentsch et al., 2008). TMMs diverge from other forms of team cognition in two fundamental ways.

First, TMMs subsume greater breadth of content, including taskwork (what the team must do in order to complete goals), teamwork (who team members interact with and how they work together collectively), and more recently, "timework" (when members interact with each other) (e.g., Cooke, Kiekel, Helm, & Usability, 2001; Mohammed et al., 2015). In contrast, other forms of team cognition tend to focus on only one category of content, such as taskwork (transactive memory systems) or teamwork (group learning) (Mohammed et al., 2010).

Second, TMM measurement requires that both content and structure be assessed. Structure refers to the relationships between concepts in team members' heads

(Mohammed et al., 2010). In contrast, other types of team cognition such as transactive memory (Lewis, 2003) or group learning (Edmondson, 1999) capture content only using Likert scales. Alternatively, TMMs assess structure via techniques such as concept mapping, in which participants determine the placement of concepts hierarchically (Marks, Sabella, Burke, & Zaccaro, 2002).

A meta-analysis concluded that TMM similarity is an important contributor to team performance outcomes such as work quality, volume, efficiency, and timeliness, above and beyond other behavioral or motivational states and team processes (DeChurch & Mesmer-Magnus, 2010a). TMMs also positively predict implicit team coordination (Rico, Sanchez-Manzanares, Gil, & Gibson, 2008) and adaptation (Burke, Stagl, Salas, Pierce, & Kendall, 2006).

Most of the TMM literature has focused on taskwork and teamwork content, but increasing attention is being given to "timework," or temporal TMMs. A temporal TMM is defined as "agreement among group members concerning deadlines for task completion, the pacing or speed of activities, and the sequencing of tasks" (Mohammed et al., 2015, p. 696). Deadlines indicate the specific time by which a task should be completed (Blount & Janicik, 2002). Pacing refers to how team members distribute effort towards completing a task over time (Blount & Janicik, 2002). Sequencing describes when steps must be completed in a specific order. In summary, team members must be on the same page regarding deadlines and pacing, as well as the ordering of subsequent steps to have high temporal TMM similarity (Mohammed et al., 2015). Initial studies found direct or conditional positive effects for temporal TMMs on team performance (Mohammed et al., 2015; Santos, Uitdewilligen, & Passos, 2015; Santos, Passos, Uitdewilligen, & Nübold, 2016). Given that the extant research on temporal TMMs is nascent, but promising, we focus on temporal TMMs in the current research.

Actual and Perceived Sharedness

Two forms of collective team cognition are structured and perceptual (DeChurch & Mesmer-Magnus, 2010a). Structured cognition captures the actual organization, patterns, and arrangement of the team's knowledge, as measured by concept maps or similarity ratings (Mohammed et al., 2010). Perceptual cognition consists of general attitudes, beliefs, values, and expectations of team members. To measure the team's awareness of their TMM similarity, team members respond to Likert scale items addressing how similar they feel components of the TMM are or the extent to which the team is on the same page (Rentsch et al., 2008). A sample item assessing perceived temporal cognition is "team members had similar thoughts about the best way to use the time available" (Gevers, Rutte, & van Eerde, 2006). Because of the measurement emphasis on structure and objectively comparing whether team members' mental maps are similar, perceptions of sharedness have received little attention in the TMM literature.

Disagreement remains regarding whether demonstrating actual similarity is enough to qualify as a shared mental model, or whether team members also have to

	Actual	
Perceived	***Not In Sync***	***In Sync***
In Sync	**False Positive** Viability High Performance Low	**True Positive** Viability Higher Performance Higher
Not In Sync	**True Negative** Viability Lower Performance Lower	**False Negative** Viability Low Performance High

FIGURE 7.1 The interactive effects of perceived and actual temporal TMM similarity on viability.

Note: High actual temporal TMM slope is significant ($p < .001$) while low actual temporal TMM slope is insignificant ($p > .05$).

perceive that they are in agreement (Mohammed et al., 2010). For example, Klimoski and Mohammed (1994) indicated that one component of a TMM is its reflection of internalized perceptions and assumptions, meaning that teams must have an understanding that they are on the same page. Conversely, Rentsch, Delise, and Hutchison (2009) argue that a team may have higher similarity of knowledge organization, qualifying as a TMM, but be ignorant of consensus. Unfortunately, it is difficult to test these competing views when few studies examine both the actual and the perceived sharedness concurrently.

Although our study variables are continuous, it is useful to note that there are four resulting categories when conceptualizing perceived and actual temporal TMM sharedness as a 2 × 2 matrix, as seen in Figure 7.1.

One category is a true positive in which a team believes it is in sync and is in actuality, and the opposite is a true negative in which a team does not perceive being in sync and is not in reality. A false positive occurs when a team perceives that it is in sync, but is not in actuality. Finally, a false negative occurs when a team perceives that it is not in sync, but actually is. Mismatches between perceptions of whether teams are in or out of sync may have differential influences on team outcomes.

The TMM literature is deficient in research examining the relationship between actual and perceived TMM similarity (Mohammed et al., 2010), and this deficiency is particularly salient in a distributed cognition context. Due to the additional difficulty sensing team members' contexts and motives (Bazarova & Walther, 2009) as well as the lack of visual and social cues (Rentsch et al., 2014), it is likely that virtual teams will struggle with temporal misperceptions and mismatches between actual and perceived TMM similarity. In addition, discrepancies may be particularly heightened for temporal TMMs, as time may be discussed less explicitly than what must be done or how it should be done in order to perform a task (Mohammed et al., 2015).

HYPOTHESES

As delineated by Hackman (1987), two forms of effectiveness examined in this study were team viability and performance. Each will be discussed below.

Viability

Main Effect

In contrast to objectively measured team performance, viability is a perceptual outcome assessing whether team members believe that they should continue to interact together in the future as a team. Viability often derives from optimistic emotions in combination with positive team member interactions (Costa, Passos, & Barata, 2015). Beyond these affective predictors, it is expected that emergent states such as TMM similarity would likely also influence viability.

When there is synchrony between team members, higher viability is expected because team members feel comfortable and compatible with one another (DePaulo & Bell, 1990). Perceptions of synchrony, through unconscious mimicry (the tendency to adopt mannerisms of another without intent), results in increased reports of rapport and liking (Lakin, Jefferis, Cheng, & Chartrand, 2003). As interpersonal synchrony results in higher perceptions of smooth interactions and rapport, team members will feel understood by other team members and other team members will feel understood by each other (DePaulo & Bell, 1990).

Blount and Janicik (2002) noted that people have an "in-sync preference," which refers to a consistent desire to be synchronous and cohesive with others. Whereas an "in-sync preference" results in favorable reactions through feeling attuned with team members, an "out of sync effect" describes the negative emotions resulting from feeling uncoordinated with others. Not only will positive feelings not occur when team members are not synchronous, but negative emotions will arise as an outcome. Such negative perceptions may decrease viability within teams (Balkundi, Barsness, & Michael, 2009). In contrast, as team members perceive each other to be in sync, they are more likely to experience rapport, liking, and smooth interactions, increasing the likelihood that they would want to work together in the future. Indeed, Rentsch and Klimoski (2001) found that agreement on teamwork TMMs was positively related to team viability. Extending this finding to temporal TMMs, we expect the following:

Hypothesis 1: Perceived temporal TMM similarity will have a positive effect on team viability.

Comparative Hypothesis

The bandwidth-fidelity argument describes the importance of measuring specific or narrow criterion with specific or narrow measures, and general or broad criterion with general or broad measures. Judge and Kammeyer-Mueller (2012) note that theoretically, as well as empirically, specific predictors are more foretelling of narrow criteria, and general predictors are more favored by broad criteria. Analogous to the bandwidth-fidelity argument, we expected that perceived TMMs are more likely to positively affect perceptual outcomes. That is, compared to actual temporal TMMs, perceived temporal TMMs would be more likely to predict viability as a perceptual criterion.

In addition to the alignment between perceptual predictors and perceptual outcomes, there is a close conceptual match between perceived temporal TMMs and viability. The team's subjective perception of having a common understanding of deadlines, ordering, and pacing is likely to positively influence their desire to work together on future projects. Because the actual temporal TMM similarity is more directly related to the objective performance scores (DeChurch & Mesmer-Magnus, 2010a, 2010b), it is anticipated that the actual similarity will be less positively related to viability than the perceived temporal TMM similarity.

> *Hypothesis 2: Compared to actual temporal TMM similarity, perceived temporal TMM similarity will have a more positive effect on team viability.*

Moderated Hypothesis

In addition to perceived temporal TMM similarity predicting viability, it is also expected that actual temporal TMM similarity will moderate this relationship. When actual temporal TMM similarity is higher, the positive relationship between the perceived temporal TMM similarity and viability is expected to increase. Team members can be more confident that they will continue to work well together in the future when actually being in sync supports perceptions of being in sync. Greater actual temporal TMM sharedness allows the team to understand and anticipate actions more successfully (Gevers et al., 2006), enhancing in sync sentiments and subsequent rapport and fondness.

However, when actual temporal TMM similarity is lower, the relationship between perceived temporal TMM similarity and viability is expected to be less positive. When teams have lower temporal TMM sharedness, confusion may result, as team actions are less likely to be foreseeable (Balkundi et al., 2009). As hints are received that team members are not on the same temporal page and tasks are being completed in an unexpected order, perceptions of confusion and uncertainty are confirmed. The realization that members are not actually in sync may discourage members from interacting or confronting these underlying conflicts, creating doubt as to whether team members will desire to work together in the future. Thus, lower actual temporal TMM similarity would strengthen the relationship of lower perceived temporal TMM similarity and lower viability.

Taking a different perspective, affective-cognitive consistency impacts the strength of job attitudes, such that when affective-cognitive consistency is low there is less strength in the relationship (Schleicher, Watt, & Greguras, 2004). Following this logic, if the actual temporal TMM (analogous to the cognitive component of the attitude) is not consistent or in sync with the perceptual (analogous to the affective component of the attitude), the relationship between actual temporal TMM similarity and viability will not be as positive as when they are consistent. Therefore, in addition to a main effect of perceived temporal TMM similarity on viability, considering the interaction between perceived and actual temporal TMM similarity allows for a more sophisticated understanding. Overall, particularly in virtual contexts, teams that believe that they are in sync, and in actuality really are in sync, will have higher viability than teams that believe that they are in sync but actually are not.

Hypothesis 3: The relationship between perceived temporal TMM similarity and viability is moderated by actual temporal TMM similarity, such that the relationship will be more positive when actual temporal TMM similarity is higher than lower.

Team Performance

Main Effect

Team performance is the most studied effectiveness outcome in the team literature (Mohammed et al., 2010). Across a variety of performance settings and outcomes, TMM similarity holds a positive relationship with team performance (Cooke et al., 2001; Edwards, Day, Arthur, & Bell, 2006; Ellis, 2006; Mathieu, Heffner, Goodwin, Cannon-Bowers, & Salas, 2005; Rentsch & Klimoski, 2001). In addition, two meta-analyses have confirmed that TMMs positively predict team performance (DeChurch & Mesmer-Magnus, 2010a, 2010b). This association is particularly true for interdependent tasks, which require additional levels of coordination of member input and effort (Leroy, Shipp, Blount, & Licht, 2015; McGrath, 1991). Temporal TMMs have also been found to be positively related to team performance (e.g., Mohammed et al., 2015). Expecting to replicate these findings, we predict:

Hypothesis 4: Actual temporal TMM similarity will have a positive effect on team performance.

Comparative Hypothesis

Drawing again from the bandwidth-fidelity argument (Judge & Kammeyer-Mueller, 2012), we suggest that actual temporal TMM similarity will relate more to performance than the perceived similarity. Compared to perceived temporal TMMs, we expect that actual temporal TMMs will be more likely to predict performance as an objective criterion. Actual temporal TMM similarity as measured through objective methods is positively related to performance outcomes in a variety of settings (DeChurch & Mesmer-Magnus, 2010a, 2010b, Mohammed et al., 2015). However, as a more subjective measure, perceived temporal TMM similarity is likely to be less related to objective performance. Perceptions are not necessarily indicative of actual similarity, and as such would not hold as positive of a relationship with performance as actual temporal TMM similarity.

Hypothesis 5: Compared to perceived temporal TMM similarity, actual temporal TMM similarity will have a more positive effect on team performance.

Moderated Hypothesis

Although actual temporal TMM similarity is anticipated to positively affect team performance, considering perceived temporal TMM similarity as a moderator is expected to affect the relationship between actually being in sync and team performance. When perceived temporal TMM similarity is higher, the relationship between actual temporal TMM similarity and team performance is expected to be more positive. When teams interpret that they are performing tasks in the expected order or working at the anticipated pace, reports of rapport, understanding, and communication increase (DePaulo & Bell, 1990; Lakin et al., 2003). This leads to better coordination of team actions, ultimately positively impacting performance outcomes (Blickensderfer, Cannon-Bowers, & Salas, 1998; Cannon-Bowers, Tannenbaum, Salas, & Volpe, 1995). In other words, this increase in communication and liking encourages discussion, including of temporal elements, resulting in higher performance by leveraging the actual temporal TMM similarity.

When teams do not perceive that they are in sync, the relationship between the actual temporal TMM similarity and performance will be less positive. Although teams may have higher actual temporal TMM sharedness, if teams perceive lower similarity, they are less likely to feel encouraged to communicate more frequently with each other to increase performance outcomes. That is, it is the perception of being out of sync and the desire to get back into sync that is likely to prompt communication to become in sync, thereby improving performance.

Just as high affective-cognitive consistency results in a more positive relationship between attitudes and performance, high levels of consistency between the actual and perceived temporal TMM sharedness will result in a more positive relationship with team performance when sharedness is high than if the perceived temporal TMM differs from the actual state.

> *Hypothesis 6: The relationship between actual temporal TMM similarity and performance is moderated by perceived temporal TMM similarity, such that the relationship will be more positive when perceived temporal TMM similarity is higher than lower.*

METHOD

Participants

The study consisted of 546 undergraduate participants from a large mid-Atlantic university, who were randomly assigned to teams consisting of three members (resulting in 182 teams). The mean age of participants was 20.15 years (*SD* = 1.01), with 69% being Caucasian and 57% being female. Participants were primarily in their first, second, or third year of attending the university. Course credit or extra credit was received in return for participation.

NeoCITIES Simulation

Students participated in NeoCITIES, an emergency crisis management team computer simulation in which participants had to respond to a range of disaster situations (McNeese et al., 2005). Each team member was randomly assigned to one of three roles: fire/emergency medical services (EMS), police, or hazardous materials (hazmat). Each role had several resources to allocate. For instance, fire/EMS had trucks, ambulances, and fire investigators to utilize. Emergency situations of varying types and levels arose, and participants were tasked with coordinating with the other team members to determine the severity of the event, sending the appropriate type and amount of resources at the correct time, and determining what resources other team members needed to address the situation. Resources were limited, and sending excess resources to a minor event could cause a delay in response to a more serious emergency due to the time required to recall resources (Hellar & McNeese, 2010). Participants were seated at a personal computer and separated by dividers. They communicated exclusively via instant messaging in a chat box.

The environment in the simulation changed depending on the students' responses, meaning that if a minor situation was not dealt with quickly, it could escalate into a larger issue. As each member received unique information, it was critical that the team communicate in order to successfully solve interdependent tasks. This hidden profile situation, in which one team member held specific knowledge about how a task should be handled, increased the interdependence between team members.

There were several advantages to using the NeoCITIES simulation in this study. The emergency crisis setting allowed for a realistic and complex domain requiring teams to coordinate and work interdependently in a virtual context. Furthermore, the simulation was designed to study team cognition and team performance (McNeese et al., 2005), with dynamic and uncertain situations. Team members also held different roles, so cooperation and timeliness of information sharing as well as decision making were essential elements for success.

Procedure

The current study was part of a larger data collection supported by the Office of Naval Research (Grant Number N000140810887; Mohammed, Hamilton, Sanchez-Manzanares, & Rico, 2017). The larger data collection included three manipulated variables: individual reflexivity (reflecting individually on one's performance), group reflexivity (reflecting as a group on the team's performance), and storytelling (conveying audio and visual information about the need for team collaboration and timing in an engaging manner using the principles of narrative). Although not of substantive interest in this study, these interventions were examined for their potential influences on team cognition (see Mohammed et al., 2017 for an overview) and controlled for in analyses.

Participants were randomly assigned to their role on the team (fire, police, hazmat) when they entered the lab. After completing an electronic survey measuring demographic information, participants were instructed on how to play the NeoCITIES simulation. Participants practiced individually for five minutes to understand the basic simulation elements and then practiced for five more minutes with other team members on a more complex, interdependent scenario. After reviewing how they performed on the training scenarios and the solutions to each event, they played two performance simulation rounds lasting 15 minutes each. The storytelling, individual reflectivity, and team reflexivity manipulations occurred following the first performance round in the order listed. After each round, teams were given feedback on how well they performed together as a team and completed an online survey, including perceived temporal TMMs, actual TMMs, and viability. Throughout this paper, measures completed after the first performance round will be referred to as Time 1 and measures completed after the second performance round will be referred to as Time 2. The total study lasted about two and a half hours.

Measures

Perceived temporal TMM similarity (Time 1 α = .86, Time 2 α = .90) was collected through responses to six survey items. Items were patterned after Gevers and colleagues (2006) and adapted to the NeoCITIES simulation to measure perceptions of deadlines, pacing, and sequencing. Sample items included, "In our team, we had the same opinion about when to arrive at certain events" and "In our team, we were on the same page regarding the deadline in which multiple units needed to arrive at events." Aggregation to the team level was justified at Time 2 (ICC(1) = .23, ICC(2) = .57, mean r_{wg} = .90). Although there is less justification at Time 1 (ICC(1) = .05, ICC(2) = .14, mean r_{wg} = .94), it is included in analyses for a more comprehensive model but of less interest.

Actual temporal TMM similarity was measured through a popular TMM measurement tool called concept mapping (Marks, Zaccaro, & Mathieu, 2000) at Times 1 and 2. Participants individually filled in three boxes (one for each role: fire, police, hazmat) indicating the sequence of the units for a given event, meaning which role should respond first, second, then third (Mohammed et al., 2015). For each dyad, one point was awarded for each shared link, and the number of shared links across the three dyads was summed to yield a team score. Higher scores indicated more shared responses (Mohammed et al., 2015).

Team viability (α = .86) was collected through four survey items derived from Tekleab, Quigley, and Tesluk (2009). After the second performance round, participants indicated their level of agreement from 1 (strongly disagree) to 5 (strongly agree) for each of the items. A sample item included "I would be happy to work with the team members on other projects in the future." Aggregation to the team-level was justified (ICC(1) = .25, ICC(2) = .50, mean r_{wg} = .88).

Team performance was measured objectively via the NeoCITIES simulation at Time 2. Because temporality is of particular interest to this study, timeliness in completing interdependent tasks was used as the performance measure. The NeoCITIES simulation calculated the average duration in seconds it took team members to complete tasks, and this value was inverted so that the average duration represented timeliness in inverted seconds (shorter average duration parallels higher timeliness; D'Innocenzo, Mathieu, & Kukenberger, 2016; Kellermanns, Walter, Lechner, & Floyd, 2005). As NeoCITIES is an emergency crisis management simulation in which a speedy response time is more ideal, higher timeliness corresponds to better team performance.

Controls: Several variables that are potentially related to TMMs and performance were controlled for in this study. Manipulations of individual reflexivity, group reflexivity, and storytelling were included as controls. As cognitive ability is positively predictive of team performance (Bell, 2007), mean team GPA was controlled for. The percentage of females on the team was also a control variable because gender composition affects performance (Baugh & Graen, 1997). Previous experience has been found to be a positive predictor of TMMs (Rentsch & Klimoski, 2001), so virtual experience (self-reported experience in a virtual environment) was controlled for. To consider the possible performance advantage that may result from previous exposure to emergency situations (e.g., prior EMT training), knowledge of emergency response protocols was also included as a control variable.

RESULTS

Descriptive Statistics

As shown in Table 7.1, actual temporal TMM similarity at Time 1 was positively related to actual similarity at Time 2 ($r = .20, p < .01$).

Similarly, perceived temporal TMM similarity at Time 1 was positively related to perceived similarity at Time 2 ($r = .32\ p < .01$). As predicted, actual temporal TMM similarity at Time 2 was positively related to timeliness ($r = .19, p < .01$). However, this relationship was not significant for actual temporal TMM similarity at Time 1. Interestingly, perceived temporal TMM similarity at Time 2 was also correlated with timeliness ($r = -.36, p < .01$), but negatively. This relationship was not significant for perceived temporal TMM similarity at Time 1. As expected, perceived temporal TMM similarity was positively related to viability at Time 1 ($r = .38, p < .01$) and 2 ($r = .51, p < .01$). Interestingly, higher timeliness was associated with lower viability ($r = -.30, p < .01$).

Data Analysis

Hierarchical regression was used to test the hypotheses at the team level. When team performance was the dependent variable, all control variables listed above were entered in Step 1. For viability as the dependent variable, the controls excluded mean GPA and knowledge of emergency response protocols because there was not compelling rationale or prior findings to include them. In Step 2, actual temporal TMM similarity at Time 1 and Time 2 were entered. Step 3 added the perceived temporal TMM similarity at Times 1 and 2. The interaction between perceived temporal TMM similarity at Time 1 and actual temporal TMM similarity at Time 1 and Time 2 were entered in Steps 4 and 5, respectively. The parallel interactions with perceived temporal TMM similarity at Time 2 were entered in Steps 6 and 7.

Tests of Hypotheses

As shown in Table 7.2, perceived temporal TMM similarity significantly and positively influenced team viability at Times 1 and 2 ($\beta = .23, p < .01$; $\beta = .43, p < .01$), lending support to Hypothesis 1.

Perceived temporal TMM similarity explained an additional 29% of unique variance in viability ($\Delta R^2 = 0.29, p < .01$) beyond controls and actual temporal TMM similarity.

According to Hypothesis 2, perceived temporal TMM similarity would have a more positive effect on viability than actual temporal TMM similarity. As predicted, perceived temporal TMM similarity at Time 1 had a stronger positive relationship with viability ($r = 0.38, p < .01$) than actual temporal TMM similarity at Time 1 ($r = -.10, p > .05$) and Time 2 ($r = -.05, p > .05$). These correlations were significantly

TABLE 7.1
Descriptive Statistics and Correlations

	M(SD)	1	2	3	4	5	6	7	8	9	10	11	12
Controls													
1. Group reflexivity condition[a]	.34(.95)	1.0											
2. Individual reflexivity condition[a]	.32(.95)	−.51**	1.0										
3. Storytelling condition[a]	−.05(1.00)	.04	.01	1.0									
4. Team percent of females[b]	0.58(.32)	.13	−.03	.18*	1.0								
5. Mean team GPA	3.23(.28)	−.03	.03	−.12	.14	1.0							
6. Experience in virtual environment	1.89(.56)	.18*	<.01	−.08	−.16*	.06	1.0						
7. Knowledge of emergency response protocols	1.66(.46)	.01	<.01	.05	−.11	−.14	.11	1.0					
Actual Temporal Similarity													
8. Actual temporal TMM similarity T1	.77(.31)	.09	−.10	.05	.21**	.08	−.08	−.04	1.0				
9. Actual temporal TMM similarity T2	.90(.32)	.18*	−.14	−.01	−.03	<.01	.06	.02	.20**	1.0			
Perceived Temporal Similarity													
10. Perceived temporal similarity T1	3.08(.43)	.01	.06	.06	−.26**	−.05	.12	.00	−.15*	−.01	1.0		
11. Perceived temporal similarity T2	3.78(.46)	.12	−.08	−.03	−.07	.10	.08	−.01	.02	.01	.32**	1.0	
Outcomes													
12. Timeliness	−57.55(11.94)	−.05	.10	.13	.15*	.02	.06	.04	.04	.19**	−.09	−.36**	1.0
13. Viability	3.86(.55)	.12	−.03	.03	−.08	−.03	.09	−.07	−.10	−.05	.38**	.51**	−.30**

Note: $N = 182$ teams.

[a] Contrast coded variable: control = −1; Group reflexivity/individual reflexivity/storytelling = 1

[b] Team gender composition of team, percent of females ranging from 0 (all males) to 1.00 (all females)

$^{*}p < .05$, two-tailed; $^{**}p < .01$, two-tailed

TABLE 7.2
Hierarchical Regression Analyses Testing the Effect of Actual and Perceived Temporal TMM Similarity on Viability

	Step 1			Step 2			Step 3			Step 4			Step 5			Step 6			Step 7		
F	1.01			1.05			9.43**			8.44**			7.64**			7.50**			6.90**		
R^2	.03			.04			.33**			.33**			.33**			.35**			.35**		
ΔR^2	.03			.01			.29**			.00			.00			.02*			.00		
Variable	*b*	*SE*	*β*	*b*	*SE*	*β*	*b*	*SE*	*β*	*b*	*SE*	*β*	*b*	*SE*	*β*	*b*	*SE*	*β*	*b*	*SE*	*β*
Step 1: Controls																					
Individual Reflexivity Condition	.02	.05	.03	.01	.05	.02	.02	.04	.01	.02	.04	.02	.01	.04	.02	.01	.04	.02	.01	.04	.02
Group Reflexivity Condition	.08	.05	.13	.09	.05	.15	.05	.04	.08	.05	.05	.08	.05	.05	.08	.05	.05	.08	.05	.05	.08
Storytelling Condition	.02	.04	.04	.02	.04	.04	.01	.04	.02	.02	.04	.03	.02	.04	.03	.02	.04	.03	.02	.04	.03
Gender Composition	−.16	.13	−.09	−.14	.14	−.08	.02	.12	.01	.03	.12	.02	.03	.12	.02	.03	.12	.02	.03	.12	.02
Experience in a Virtual Environment	.05	.08	.06	.05	.08	.05	.02	.07	.02	.02	.07	.02	.02	.07	.02	.02	.07	.02	.02	.07	.02
Step 2: Actual Similarity																					
Actual Temporal TMM Similarity T1				−.14	.14	−.08	−.14	.12	−.08	−.10	.12	−.06	−.13	.12	−.06	−.11	.12	−.06	−.11	.12	−.06
Actual Temporal TMM Similarity T2				−.11	.13	−.07	−.10	.11	−.06	−.11	.11	−.07	−.10	.11	−.07	−.11	.11	−.07	−.10	.11	−.07
Step 3: Perceived Similarity																					
Perceived Temporal TMM Similarity T1							.30	.09	.23**	.30	.09	.23**	.30	.09	.23**	.30	.09	.23**	.30	.09	.23**
Perceived Temporal TMM Similarity T2							.52	.08	.43**	.51	.08	.43**	.52	.08	.43**	.52	.08	.43**	.52	.08	.43**
Step 4: Actual T1 × Perceived T1										.03	.28	.01	.03	.28	.01	−.14	.29	−.03	−.14	.29	−.03
Step 5: Actual T2 × Perceived T1													.06	.27	.01	.03	.26	.01	.07	.26	.01
Step 6: Actual T1 × Perceived T2																.49	.23	.14*	.50	.23	.14*
Step 7: Actual T2 × Perceived T2																			−.10	.30	−.02

Note: $N = 182$ teams
$^*p < .05$, $^{**}p < .01$

different at both Times 1 and 2, respectively (Z = 4.35, $p < .01$; Z = 4.16, $p < .01$; Meng, Rosenthal, & Rubin, 1992). Similarly, perceived temporal TMM similarity at Time 2 had a stronger positive relationship with viability ($r = 0.51, p < .01$) than actual temporal TMM similarity at Times 1 and 2. These correlations were also significantly different at both Times 1 and 2 (Z = −6.13, $p < .01$; Z = −5.65, $p < .01$), supporting Hypothesis 2.

Hypothesis 3 predicted that the relationship between perceived temporal TMM similarity and viability would be moderated by actual temporal TMM similarity, such that the relationship will be more positive when actual temporal TMM similarity is higher than lower. As shown in Table 7.2, actual temporal TMM similarity at Time 1 significantly moderated the effect of perceived similarity (at Time 2) on viability ($\beta = .14$, $p < .05$). The interaction explained an additional 2% of unique variance in viability ($\Delta R^2 = 0.02$, $p < .05$) beyond controls and main effects. As recommended by Aiken, West, and Reno (1991), the individual regression slopes were graphed. Figure 7.2 shows that the slope of teams with high actual temporal TMM similarity portrayed a significant, positive relationship ($t = 4.37$, $p < .001$).

Meanwhile, the positive relationship of teams with low actual similarity was insignificant ($t = .59$, $p > .05$). When perceptions of similarity were high, teams with high actual similarity trended in the direction of higher viability than those with low actual similarity, but not significantly so ($t = .46$, $p > .05$). However, when perceptions of similarity were low, ratings of viability were lower when teams had high actual similarity than when teams had low actual similarity ($t = -2.43$, $p < .05$). The interaction between perceived (at Time 2) and actual temporal TMM similarity at Time 2 on viability was not significant, nor were the interactions between actual (Time 1 or 2) and perceived at Time 1. This pattern of results lends support to Hypothesis 3 only for actual similarity at Time 1 and perceived similarity at Time 2.

Regarding Hypothesis 4 as seen in Table 7.3, actual similarity at Time 1 did not significantly relate to timeliness ($\beta = -.01$, $p > .05$).

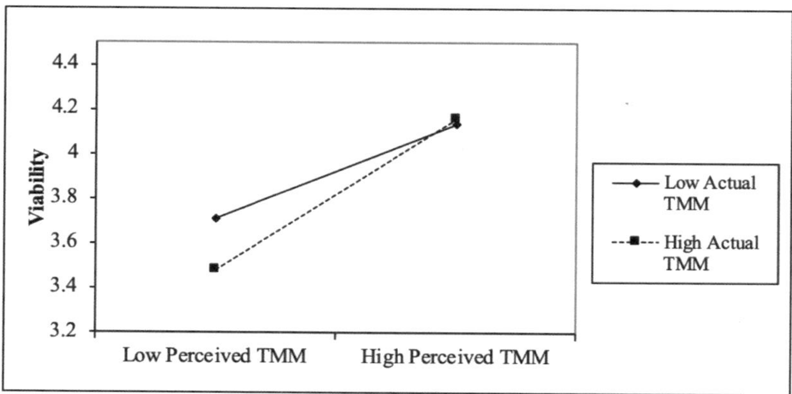

FIGURE 7.2 The interactive effects of perceived and actual temporal TMM similarity on viability.

Note: High actual temporal TMM slope is significant ($p < .001$) while low actual temporal TMM slope is insignificant ($p > .05$).

TABLE 7.3
Hierarchical Regression Analyses Testing the Effect of Actual and Perceived Temporal TMM Similarity on Performance

	Step 1			Step 2			Step 3			Step 4			Step 5			Step 6			Step 7		
F	1.47			2.19*			4.28**			4.26**			3.95**			3.65**			3.81**		
R^2	.06			.10*			.22**			.23**			.23**			.23**			.26**		
ΔR^2	.06			.05*			.11**			.02			.00			.00			.02*		
Variable	*b*	*SE*	*β*	*b*	*SE*	*β*	*b*	*SE*	*β*	*b*	*SE*	*β*	*b*	*SE*	*β*	*b*	*SE*	*β*	*b*	*SE*	*β*
Step 1: Controls																					
Individual Reflexivity Condition	.87	1.09	.07	1.04	1.07	.08	.91	1.00	.07	1.03	1.00	.08	1.02	1.00	.08	1.02	1.00	.08	1.18	1.00	.09
Group Reflexivity Condition	–.82	1.13	–.07	–1.22	1.11	–.10	–.70	1.05	–.06	–.76	1.04	–.06	–.81	1.05	–.06	–.80	1.05	–.06	–.49	1.05	–.04
Storytelling Condition	1.23	.91	.10	1.24	.89	.10	1.18	.84	.10	1.12	.84	.10	1.18	.84	.10	1.19	.85	.10	1.18	.84	.11
Gender Composition	6.19	2.93	.17*	6.65	2.92	.18*	5.79	2.84	.16*	5.85	2.82	.16*	5.83	2.82	.16*	5.84	2.83	.16*	5.77	2.80	.15*
Experience in a Virtual Environment	2.14	1.65	.10	2.00	1.62	.09	2.23	1.53	.11	2.01	1.52	.10	2.06	1.52	.10	2.07	1.53	.10	2.58	1.53	.12
Mean Team GPA	–1.68	3.21	–.04	–1.76	3.15	–.04	–.08	2.98	–.01	–.29	2.96	–.01	–.25	2.97	–.01	–.19	2.99	–.01	–.26	2.96	–.01
Knowledge of Emergency Response	.82	1.95	.03	.77	1.91	.03	.78	1.80	.03	.89	1.78	.03	.97	1.79	.04	.97	1.80	.04	1.00	1.78	.04
Step 2: Actual Similarity																					
Actual Temporal TMM Similarity T1				–.57	2.91	–.02	–.06	2.75	–.01	–.89	2.77	–.02	–1.04	2.78	–.03	–.99	2.80	–.03	–.97	2.77	–.03
Actual Temporal TMM Similarity T2				8.31	2.80	.22**	8.00	2.63	.22**	8.67	2.64	.23**	8.74	2.64	.24**	8.71	2.65	.23**	8.11	2.64	.22**
Step 3: Perceived Similarity																					
Perceived Temporal TMM Similarity T1							1.29	2.12	.05	1.58	2.11	.06	1.54	2.11	.06	1.54	2.12	.06	1.64	2.09	.06
Perceived Temporal TMM Similarity T2							–9.30	1.91	–.36**	–9.13	1.90	–.35**	–9.16	1.90	–.35**	–9.16	1.91	–.35**	–10.23	1.95	–.39**
Step 4: Actual T1 × Perceived T1										–12.18	6.56	–.13	–12.27	6.57	–.13	–12.67	6.85	–.14	–12.41	6.78	–.13
Step 5: Actual T2 × Perceived T1													3.84	6.21	.04	3.78	6.23	.04	–2.80	6.84	–.03
Step 6: Actual T1 × Perceived T2																1.20	5.58	.02	–.47	5.57	–.01
Step 7: Actual T2 × Perceived T2																			15.27	6.90	.18*

Note: $N = 182$ teams.
$^{*}p < .05$, $^{**}p < .01$

However, actual similarity at Time 2 was associated with increased timeliness ($\beta = .20$, $p < .01$). The actual temporal TMM similarity at Times 1 and 2 explained a significant 5% of unique additional variance in the timeliness beyond controls and perceived similarity ($\Delta R^2 = 0.05$, $p < .01$). Hypothesis 4 was supported for actual temporal TMM similarity measured at Time 2, but not at Time 1.

Hypothesis 5 stated that actual temporal TMM similarity would have a stronger positive relationship with performance than perceived temporal TMM similarity. Although actual temporal TMM similarity had a positive relationship with timeliness at Time 2 as expected ($r = 0.19$, $p < .01$), the correlation magnitude was stronger for perceived temporal TMM similarity at Time 2 and in the opposite direction than predicted ($r = -.36$, $p < .01$; $Z = -5.30$, $p < .01$). There was no significant relationship between actual or perceived temporal TMM similarity at Time 1 with timeliness. Hypothesis 5 was not supported.

Hypothesis 6 predicted that perceived temporal TMM similarity would moderate the relationship between actual temporal TMM similarity and performance, such that the relationship would be more positive when perceived temporal TMM similarity is higher than lower. Perceived similarity at Time 2 significantly moderated the effect of actual similarity at Time 2 on timeliness ($\beta = .18$, $p < .05$). The interaction explained an additional 2% unique variance in performance ($\Delta R^2 = 0.02$, $p < .05$) beyond controls and main effects. A graph of the interaction (see Figure 7.3) demonstrated that the slope of teams with higher perceived temporal TMM similarity portrayed a significant, positive relationship ($t = 3.31$, $p = .001$) with timeliness, as expected.

Meanwhile, the positive relationship of teams with low perceived similarity and timeliness was insignificant ($t = -.95$, $p > .05$). When teams had high actual

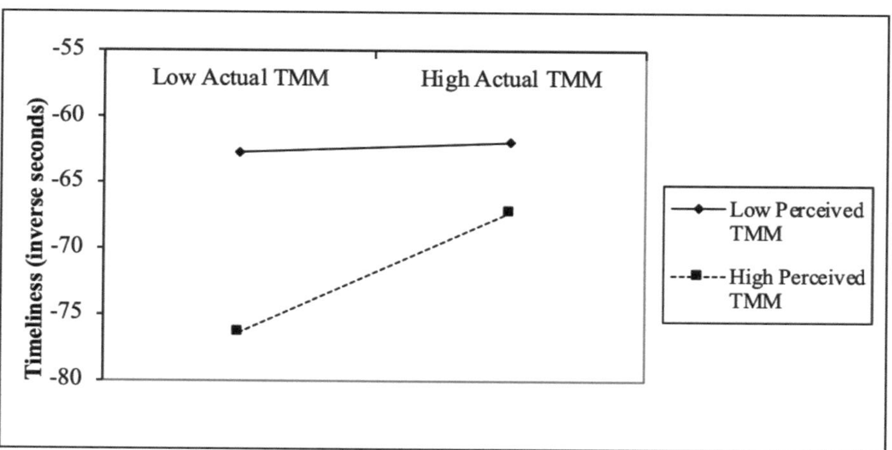

FIGURE 7.3 The interactive effects of actual and perceived temporal TMM similarity on performance.

Note: High perceived temporal TMM slope is significant ($p < .01$) while low perceived temporal TMM slope is insignificant ($p > .05$).

similarity, those with high perceived similarity were slightly less timely than those with low perceived similarity ($t = 2.37, p < .05$). However, when teams had low actual similarity, those with high perceived similarity were much less timely than those with low perceived similarity ($t = 4.76, p < .01$). The interaction of actual similarity at Time 1 and perceived similarity at Time 2 on timeliness was not significant. The interactions between actual (Time 1 or 2) and perceived at Time 1 were also insignificant. This pattern of results lends support to Hypothesis 6 with both perceived and actual similarity at Time 2, but not at Time 1.

DISCUSSION

This study yielded several significant findings. First, although perceived temporal TMM similarity has not received much attention to date, it exerted a stronger influence on both viability and performance than the actual temporal TMM similarity. Second, we consistently found that teams with a mismatch between actual and perceived similarity experienced lower viability and performance, even lower than those who had both low actual and low perceived similarity. Third, although our hypotheses were agnostic to time, significant differences emerged at Time 1 and Time 2. These findings are discussed below.

First, compared to actual temporal TMM similarity, perceived temporal TMM similarity exerted a stronger influence on both viability and performance. Although expected for viability, perceived TMM similarity unexpectedly influenced team performance more strongly compared to actual TMM similarity. Whereas actual TMM similarity was related to higher performance as predicted, surprisingly, higher perceived TMM similarity was associated with lower performance (less timeliness). Because timeliness was our performance measure, increased or inefficient communication may take up more time, especially in virtual teams, resulting in less timeliness. More communication within a team is typically expected to lead to more liking (DePaulo & Bell, 1990; Lakin et al., 2003) and better coordination of team actions (Cannon-Bowers et al., 1995), which improve performance (e.g., Rico et al., 2008). However, in a virtual team, increased communication may require more time, and misunderstandings may take longer to rectify, which would have a detrimental effect on the performance metric of timeliness. Indeed, virtual teams are prone to receiving and discussing too much irrelevant information, making for dysfunctional information exchange (Ellwart et al., 2015) that may result in lower timeliness. Additionally, multiple members repeating the same information signals that it is important (McLeod, 2013) and is likely to align perceptions of similarity, but may be especially time consuming in virtual teams, decreasing timeliness.

Second, study results demonstrated the importance of the match between actual and perceived temporal TMM similarity in virtual teams. Even more than having both low actual and low perceived TMM similarity, results revealed that mismatches between actual and perceived temporal TMM similarity were particularly detrimental to both viability and performance in virtual contexts. Regarding viability, teams that thought they were in sync and actually were in sync were more likely to desire working together in the future. Viability suffered the most when teams perceived low temporal similarity, but actually had high similarity. However, this relationship was only true when actual and perceived temporal TMM similarity were measured at Time 1 and Time 2, respectively.

Potentially, rapport and understanding impacted by the level of actual temporal similarity in the first round (Gevers et al., 2006) may set the stage for affecting viability, measured after the second round. Perceptions of relationships can develop early and tend to remain consistent thereafter (Liden, Wayne, & Stilwell, 1993). Kimble (2011) notes the importance of members' first "online impression" in virtual teams, as it influences conversational tones for later discussions. Initial online impressions may be impacted by the actual similarity early on and influence later perceptions of similarity. The early formation of relationship quality as influenced by initial actual similarity may theoretically explain the relationship of the actual temporal TMM similarity at Time 1 interacting with perceptions of similarity at Time 2 to impact viability.

Regarding performance, performance was lower when teams thought they were in sync, but actually were not. Similar to the findings for viability, performance was lowest when teams were actually not on the same page, but perceived they were. This relationship was only true when actual and perceived temporal TMM similarity were measured at Time 2 and not at Time 1, which is consistent with prior findings that temporal TMMs have a greater impact on performance later in a team's lifespan rather than earlier. Mohammed and colleagues (2015) found that actual temporal TMMs at Time 1 were not significantly related to performance, but temporal TMMs measured at Time 2 were positively related to performance.

Although it was inferred that teams low in both perceived and actual TMM similarity would have lowest performance, mismatches were even more detrimental. Clarkson, Hirt, Jia, and Alexander (2010) reported similar negative outcomes regarding mismatched actual versus perceived depletion. Specifically, individuals that were given feedback stating they should be less depleted (perceived) while in the high depletion condition (actual), or actually had low depletion but were told they should have high depletion, were less persistent in the task, made more errors, and had a longer response time than individuals who were told they should be depleted and were also actually depleted. Overall, teams with mismatches performed worse than those with both low actual and perceived similarity.

Implications

Although not receiving much emphasis in the TMM literature to date (Mohammed et al., 2010), perceived temporal TMM similarity exerted a stronger influence on both viability and performance compared to actual temporal TMM similarity. Team distribution complicates the extent to which members are aware of others' progress and performance (Schmidtke & Cummings, 2017), making it likely that perceived TMM similarity may differ from actuality. Indeed, including both perceived and actual temporal TMM similarity in the model accounted for additional variance in predicting viability and performance over just one or the other. Considering both, particularly the perceived similarity, appears especially important for virtual teams.

The empirical contribution of measuring both actual and perceived similarity at once allowed for an examination of interactions, which demonstrated that some teams' perceptions and actual temporal TMM similarity were in agreement whereas others' were not, as Rentsch and colleagues (2009) proposed. The results of this study corroborated that when teams are not in sync, negative outcomes on viability and performance result. There was a negative effect on viability for teams that had low perceived similarity, but high levels in actuality. For these teams, ratings of team viability were remarkably even lower than teams that had low actual similarity. This resulted in teams that were on the same temporal page, but did not desire to work together in the future due to their misperception.

Additionally, poor performance resulted for virtual teams that had low actual similarity, but perceived high similarity. Teams that rated themselves as having high perceptions of temporal similarity but demonstrated low actual similarity surprisingly were less timely than teams that had low actual similarity and also perceived low similarity.

Results demonstrated that perceived similarity had a stronger effect on viability and performance than actual similarity. Therefore, a practical implication is that perceptions of being in sync do matter and that virtual team members should be aware that perceptions of temporal TMM similarity are important. Study results also revealed the importance of having synchrony between actual and perceived temporal TMM similarity in distributed contexts. When performance or viability is low, it may be beneficial for virtual teams to explore the potential (mis)match between their actual and perceived TMM similarity.

Limitations and Future Directions

Although focusing on under-researched temporal TMMs was a contribution (Mohammed et al., 2012), results may not be generalizable to taskwork or teamwork TMMs. Future research should compare actual versus perceived TMM similarity for taskwork and teamwork content. Similarly, as the performance outcome in this study was timeliness, there are likely a number of additional temporal team inputs and processes such as temporal individual differences within the team and related process conflict (Mohammed, Hamilton, & Lim, 2009) that may also affect team timeliness. Exploring the influence of these variables would also be beneficial. Of note, generalizing the current findings to alternative forms of performance such as quality or quantity of virtual team output should be done cautiously.

A high degree of task interdependence, dynamic events, decision making, and immediate feedback in the NeoCITIES simulation made it an excellent tool to study TMM similarity in a virtual context. However, teams participating in the study only worked together for about two and a half hours, therefore results may be more applicable to newly formed teams. Future research should examine the differences between perceived and actual TMMs in more lengthy projects or long-term teams. In addition, it would be beneficial to examine the impact of TMMs in virtual teams distributed across varying geographical locations and time zones (Schmidtke & Cummings, 2017).

As the content of the concept maps was based specifically on the NeoCITIES simulation (ordering the units to respond to a particular event), it was necessary to collect these measures following the performance rounds so participants were familiar with the content (Ellis, 2006). Therefore, causality cannot be claimed. Future studies should incorporate an experimental design.

Another direction for future research is to include more measurement time points. Due to the time-consuming nature of the training required for the simulation and the added difficulty of measuring concept maps, it was feasible to only have two measurement points and performance rounds (Mohammed et al., 2015). However, it would be beneficial to develop a more thorough understanding of the nature of actual and perceived TMMs over time in virtual contexts by incorporating more time points, potentially in a long-term team setting.

Because not being in sync is detrimental to performance and viability outcomes in virtual teams, researchers should explore methods of getting in sync. One possibility is cross-training as a training technique to better align perceptions of temporal TMM similarity with actual levels. Cross-training can improve interpersonal knowledge by learning about duties of other team members, which contributes to team shared understanding (Blickensderfer et al., 1998). Therefore, cross training may prove particularly useful for developing similar TMMs in virtual teams (Schmidtke & Cummings, 2017), as this process is particularly complex in distributed contexts (Ellwart et al., 2015).

Conclusion

Despite the prevalence of TMM studies, "there is a dearth of empirical research investigating the role of shared cognition in the virtual team literature" (Schmidtke & Cummings, 2017, p. 673). Answering this call, we examined the impact of perceived and actual temporal TMM similarity on performance and viability in virtual teams. Compared to actual temporal TMM similarity, perceived temporal TMM similarity exerted a stronger influence on both viability and performance. Additionally, mismatches between actual and perceived TMM similarity were particularly detrimental to outcomes. When teams think they are temporally in sync but were actually not, performance declined. When teams perceived low levels of sharedness but held high actual similarity, viability decreased. Despite receiving little attention in the TMM literature, results highlight the importance of perceived temporal TMM similarity in addition to actual TMM similarity.

REFERENCES

Aiken, L., West, S., & Reno, R. (1991). *Multiple regression: Testing and interpreting interactions*. Newbury Park, CA: Sage Publications.

Balkundi, P., Barsness, Z., & Michael, J. (2009). Unlocking the influence of leadership network structures on team conflict and viability. *Small Group Research*, *40*(3), 301–322.

Baugh, G., & Graen, G. (1997). Effects of team gender and racial composition on perceptions of team. *Group and Organization Management*, *22*(3), 366–383.

Bazarova, N. N., & Walther, J. B. (2009). Attributions in virtual groups. *Small Group Research*, *40*(2), 138–162. https://doi.org/10.1177/1046496408328490

Bell, S. T. (2007). Deep-level composition variables as predictors of team performance: A meta-analysis. *The Journal of Applied Psychology*, *92*(3), 595–615. https://doi.org/10.1037/0021-9010.92.3.595

Blickensderfer, E., Cannon-Bowers, J., & Salas, E. (1998). Cross-training and team performance. In J. Cannon-Bowers & E. Salas (Eds.), *Making decisions under stress: Implications for individual and team training* (pp. 299–311). Washington, DC: American Psychological Association.

Blount, S., & Janicik, G. (2002). Getting and staying in pace: The "in-synch" preference and its implications for work groups. In *Research on managing groups and teams: Toward phenomenology of groups and group members* (Vol. 4, pp. 235–266). New York. Elsevier Science.

Burke, C. S., Stagl, K. C., Salas, E., Pierce, L., & Kendall, D. (2006). Understanding team adaptation: A conceptual analysis and model. *The Journal of Applied Psychology, 91*(6), 1189–1207. https://doi.org/10.1037/0021-9010.91.6.1189

Cannon-Bowers, J. A., Salas, E., & Converse, S. (1993). Shared mental models in expert team decision making. *Individual and Group Decision Making: Current Issues, 221.*

Cannon-Bowers, J. A., Tannenbaum, S. I., Salas, E., & Volpe, C. E. (1995). Defining competencies and establishing team training requirements. In R. Guzzo & E. Salas (Eds.), *Team effectiveness and decision making in organizations* (pp. 333–380). San Francisco, CA: Jossey Bass.

Clarkson, J. J., Hirt, E. R., Jia, L., & Alexander, M. B. (2010). When perception is more than reality: The effects of perceived versus actual resource depletion on self-regulatory behavior. *Journal of Personality and Social Psychology, 98*(1), 29–46. https://doi.org/10.1037/a0017539

Cooke, N. J., Kiekel, P. A., Helm, E. E., & Usability, A. (2001). Acquisition of a complex task. *International Journal, 5*(3), 297–315.

Costa, P. L., Passos, A. M., & Barata, M. C. (2015). Multilevel influences of team viability perceptions. *Team Performance Management, 21*(1/2), 19–36.

DeChurch, L. A., & Mesmer-Magnus, J. R. (2010a). Measuring shared team mental models: A meta-analysis. *Group Dynamics, 14*(1), 1–14. https://doi.org/10.1037/a0017455

DeChurch, L. A, & Mesmer-Magnus, J. R. (2010b). The cognitive underpinnings of effective teamwork: A meta-analysis. *The Journal of Applied Psychology, 95*(1), 32–53. https://doi.org/10.1037/a0017328

DePaulo, B., & Bell, K. L. (1990). Rapport is not so soft anymore. *Psychological Inquiry, 1*, 305–308. https://doi.org/10.1207/s15327965pli0104 6

D'Innocenzo, L., Mathieu, J. E., & Kukenberger, M. R. (2016). A meta-analysis of different forms of shared leadership—Team performance relations. *Journal of Management, 40*, 1–28.

Edmondson, A. (1999). Psychological safety and learning behavior in work teams. *Administrative Science Quarterly, 44*(2), 350–383. Retrieved from www.jstor.org/stable/2666999

Edwards, B. D., Day, E. A., Arthur, W., & Bell, S. T. (2006). Relationships among team ability composition, team mental models, and team performance. *Journal of Applied Psychology, 91*, 727–736.

Ellis, A. P. J. (2006). System breakdown: The role of mental models and transactive memory in the relationship between acute stress and team performance. *Academy of Management, 49*(3), 576–589.

Ellwart, T., Happ, C., Gurtner, A., & Rack, O. (2015). Managing information overload in virtual teams: Effects of a structured online team adaptation on cognition and performance. *European Journal of Work and Organizational Psychology, 24*(5), 812–826. https://doi.org/10.1080/1359432X.2014.1000873

Gevers, M., Rutte, C., & van Eerde, W. (2006). Meeting deadlines in work groups. *Applied Psychology: An International Review, 55*(1), 52–72. https://doi.org/10.1111/j.1464-0597.2006.00228.x

Gilson, L. L., Maynard, M. T., Jones Young, N. C., Vartiainen, M., & Hakonen, M. (2015). Virtual teams research: 10 years, 10 themes, and 10 opportunities. *Journal of Management*, *41*(5), 1313–1337. https://doi.org/10.1177/0149206314559946

Hackman, J. R. (1987). The design of work teams. In J. W. Lorsch (Ed.), *Handbook of organizational behavior* (pp. 315–342). Englewood Cliffs, NJ: Prentice Hall.

Hellar, D. B., & McNeese, M. (2010). NeoCITIES: A simulated command and control task environment for experimental research. In *Proceedings of the human factors and ergonomics society annual meeting* (Vol. 54, No. 13, pp. 1027–1031). Los Angeles, CA: SAGE Publications.

Hollan, J., Hutchins, E., & Kirsh, D. (2000). Distributed cognition: Toward a new foundation for human-computer interaction research. *ACM Transactions on Computer-Human Interaction*, *7*(2), 174–196. https://doi.org/10.1145/353485.353487

Judge, T. A., & Kammeyer-Mueller, J. D. (2012). General and specific measures in organizational behavior research: Considerations, examples, and recommendations for researchers. *Journal of Organizational Behavior*, *33*, 161–174.

Kellermanns, F., Walter, J., Lechner, C., & Floyd, S. (2005). The lack of consensus about strategic consensus: Advancing theory and research. *Journal of Management*, *31*(5), 719–737.

Kimble, C. (2011). Building effective virtual teams: How to overcome the problems of trust and identity in virtual teams. *Global Business and Organizational Excellence*, *30*(2), 6–15. https://doi.org/10.1002/joe.20364

Klimoski, R., & Mohammed, S. (1994). Team mental model: Construct or metaphor? *Journal of Management*, *20*, 403–437. *Journal of Management*, *31*(5), 717–737.

Lakin, J. L., Jefferis, V. E., Cheng, C. M., & Chartrand, T. L. (2003). The chameleon effect as social glue: Evidence for the evolutionary significance of nonconscious mimicry. *Journal of Nonverbal Behavior*, *27*, 145–162. https://doi.org/10.1023/A:1025389814290

Leroy, S., Shipp, A. J., Blount, S., & Licht, J. G. (2015). Synchrony preference: Why some people go with the flow and some don't. *Personnel Psychology*, *68*(4), 759–809. https://doi.org/10.1111/peps.12093

Lewis, K. (2003). Measuring transactive memory systems in the field: Scale development and validation. *Journal of Applied Psychology*, *88*, 587–604.

Liden, R. C., Wayne, S. J., & Stilwell, D. (1993). A longitudinal study on the early development of leader-member exchanges. *Journal of Applied Psychology*, *78*, 662–674.

Marks, M. A., Sabella, M. J., Burke, C. S., & Zaccaro, S. J. (2002). The impact of cross-training on team effectiveness. *Journal of Applied Psychology*, *87*, 3–13.

Marks, M. A., Zaccaro, S. J., & Mathieu, J. E. (2000). Performance implications of leader briefings and team-interaction training for team adaptation to novel environments. *Journal of Applied Psychology*, *85*, 971–986.

Mathieu, J. E., Heffner, T. S., Goodwin, G. F., Cannon-Bowers, J. A., & Salas, E. (2005). Scaling the quality of teammates' mental models: Equifinality and normative comparisons. *Journal of Organizational Behavior*, *26*(1), 37–56. https://doi.org/10.1002/job.296

Maynard, M. T., & Gilson, L. L. (2014). The role of shared mental model development in understanding virtual team effectiveness. *Group and Organization Management*, *39*(1), 3–32. https://doi.org/10.1177/1059601113475361

McGrath, J. E. (1991). Time, Interaction, and Performance (TIP): A theory of groups. *Small Group Research*, *22*(2), 147–174.

McLeod, P. L. (2013). Distributed people and distributed information: Vigilant decision-making in virtual teams. *Small Group Research*, *44*(6), 627–657. https://doi.org/10.1177/1046496413500696

McNeese, M., Bains, P., Brewer, I., Brown, C., Connors, E., Jefferson, T., . . . Terrel, I. (2005). The NeoCITIES simulation: Understanding the design and experimental methodology used to develop a team emergency management simulation. In *Proceedings of the Human Factors and Ergonomics Society Annual Meeting* (Vol. 49, No. 3, pp. 591–594). Los Angeles, CA: SAGE Publications.

Meng, X., Rosenthal, R., & Rubin, D. (1992). Comparing correlated correlation coefficients. *Psychological Bulletin, 111*(1), 172–175.

Mesmer-Magnus, J. R., DeChurch, L. A., Jimenez-Rodriguez, M., Wildman, J., & Shuffler, M. (2011). A meta-analytic investigation of virtuality and information sharing in teams. *Organizational Behavior and Human Decision Processes, 115*(2), 214–225. https://doi.org/10.1016/j.obhdp.2011.03.002

Mohammed, S., Ferzandi, L., & Hamilton, K. (2010). Metaphor no more: A 15-year review of the team mental model construct. *Journal of Management, 36*(4), 876–910. https://doi.org/10.1177/0149206309356804

Mohammed, S., Hamilton, K., & Lim, A. (2009) The incorporation of time in team research: Past, current, and future. In E. Salas, G. F. Goodwin, & C. S. Burke (Eds.), *Team effectiveness in complex organizations: Cross-disciplinary perspectives and approaches.* New York: Routledge Taylor & Francis Group.

Mohammed, S., Hamilton, K., Tesler, R., Mancuso, V., & McNeese, M. (2015). Time for temporal team mental models: Exploring beyond "what" and "how" to incorporate "when". *European Journal of Work and Organizational Psychology, 24*(5), 693–709.

Mohammed, S., Hamilton, K., Sanchez-Manzanares, M., & Rico, R. (2017). Team cognition: Team mental models and situation awareness. In E. Salas, R. Rico, & J. Passmore (Eds.), *The Wiley Blackwell handbook of the psychology of teamwork and collaborative processes.* Hoboken, NJ: John Wiley & Sons, Ltd.

Mohammed, S., McNeese, M., Mancuso, V., Hamilton, K., & Tesler, R. (2012). *The role of metaphors in fostering macrocognitive processes in distributed teams.* Final report prepared for the Office of Naval Research.

Rentsch, J. R., Delise, L. A., & Hutchison, S. (2009). Team member schema accuracy and team member schema congruence: In search of the Team MindMeld. In E. Salas, G. F. Goodwin, & C. S. Burke (Eds.), *Team effectiveness in complex organizations* (pp. 241–266). New York: Routledge, Taylor & Francis Group.

Rentsch, J. R., Delise, L. A., Mello, A. L., & Staniewicz, M. J. (2014). The integrative team knowledge building training strategy in distributed problem-solving teams. *Small Group Research, 45*(5), 568–591. https://doi.org/10.1177/1046496414537690

Rentsch, J. R., & Klimoski, R. J. (2001). Why do "great minds" think alike?: Antecedents of team member schema agreement. *Journal of Organizational Behavior, 22*, 107–120. https://doi.org/10.1002/job.81

Rentsch, J. R., Small, E. E., & Hanges, P. J. (2008). Cognitions in organizations and teams: What is the meaning of cognitive similarity? In B. Smith (Ed.), *The people make the place* (pp. 127–157). Mahwah, NJ: Lawrence Erlbaum.

Rico, R., Sanchez-Manzanares, M., Gil, F., & Gibson, C. (2008). Team implicit coordination processes: A team knowledge-based approach. *Academy of Management Review, 33*, 163–184.

Salas, E., Sims, D. E., & Burke, C. S. (2005). Is there a "big five" in teamwork? *Small Group Research, 36*, 555–599.

Santos, C. M., Passos, A. M., Uitdewilligen, S., & Nübold, A. (2016). Shared temporal cognitions as substitute for temporal leadership: An analysis of their effects on temporal conflict and team performance. *Leadership Quarterly, 27*(4), 574–587. https://doi.org/10.1016/j.leaqua.2015.12.002

Santos, C., Uitdewilligen, S., & Passos, A. (2015). A temporal common ground for learning: The moderating effect of shared mental models on the relation between team learning behaviours and performance improvement. *European Journal of Work and Organizational Psychology, 24*(5), 710–725.

Schleicher, D. J., Watt, J. D., & Greguras, G. J. (2004). Reexamining the job satisfaction–performance relationship: The complexity of attitudes. *Journal of Applied Psychology, 89*(1), 165–177.

Schmidtke, J. M., & Cummings, A. (2017). The effects of virtualness on teamwork behavioral components: The role of shared mental models. *Human Resource Management Review, 27*(4), 660–677. https://doi.org/10.1016/j.hrmr.2016.12.011

Tekleab, A. G., Quigley, N. R., & Tesluk, P. E. (2009). A longitudinal study of team conflict, conflict management, cohesion, and team effectiveness. *Group & Organization Management, 34*(2), 170–205.

Wellens, A. R. (1993). Group situation awareness and distributed decision-making: From military to civilian applications. In N. J. Castellan Jr. (Ed.), *Individual and group decision making: Current issues* (pp. 267–291). Hillsdale, NJ: Lawrence Erlbaum.

8 Expertise and Distributed Team Cognition

A Critical Review and Research Agenda

James A. Reep

CONTENTS

INTRODUCTION

Understanding expertise is an important endeavor in a variety of research contexts. Cognitive scientists seek to understand individual differences of knowledge structures and representations, and their influence on decision-making processes (Wright & Bolger, 1992). AI researchers are interested in how experts learn in hopes of creating intelligent systems that perform at high or human-like levels (Hambrick & Hoffman, 2016). Educationalists seek an understanding of how experts are created so that methods of instruction and training can be improved (Alexander, 2005; Bereiter & Scardamalia, 1986).

Within complex fields of practice, formalized education (i.e. attending a college or university) provides a much-needed knowledge structure for students (Alexander, 2005; Bereiter & Scardamalia, 1986; Goldman & Petrosino, 1999; Mehta, Suto, Elliott, & Rushton, 2011) but often lacks the specialized task and contextual knowledge employers need for them to be autonomous employees, or experts (Bishop, 1989). To mitigate the lack of expertise for these newly hired employees, informal and formal training programs can be implemented to supplement an individual's knowledge gap. The problem that arises in these programs, however, is that they neglect the teamwork aspect of the environment, which has been identified as being important to employers (Hesketh, 2000). As such, having someone familiar with the context (i.e. an expert) train alongside a novice in these settings would be beneficial in addressing teamwork and communication between novice and expert employees, emulating some of these very characteristics (i.e., learning, problem solving, communication, and others) during the training (Tracey et al., 2015; Urick, 2017).

Unfortunately, experts and novices[1] have very different understandings and perspectives on the environment (Hinds, Patterson, & Pfeffer, 2001; Hmelo-Silver & Pfeffer, 2004), how they approach problem solving (Fischer, Greiff, & Funke, 2012; Gick, 1986; Klein & Borders, 2016), and how their cognition is distributed (or not) across content and artifacts within the environment (Hollan, Hutchins, & Kirsh, 2000; Hutchins, 1995). This presents organizations with the challenge of imbuing novices with the knowledge contained within individual experts with years of service, particularly due to the disparate mental models of these two group classifications (i.e. experts versus novices) and the composition of their cognitive processes. More pointedly, experts' cognitive processes are distributed among their internal biological cognitive processes, the context, and the artifacts within their environment, whereas conversely, novices do not possess the strong connections between the environment and the task limiting the distribution of their cognitive processes.

To better understand expertise, this chapter will critically analyze the extant literature on the topic, analyze how it is measured and used, and declare a future research agenda employing the information explicated.

EXPERTISE

Because of the important role that experts play in our society, the nature of expertise has been the topic of much research. Most researchers define expertise as knowledge and experience gained from spending a significant amount of deliberate practice at a particular skill or within a particular domain (Ericsson, Krampe, & Tesch-Römer, 1993; Ericsson, Prietula, & Cokely, 2007; Hambrick et al., 2014). The key to developing expertise is this idea of deliberate practice, which "entails considerable, specific, and sustained efforts to do something you can't do well—or even at all" (Ericsson et al., 2007, p. 3).

The general amount of time ascribed for an individual becoming an expert is about 10,000 hours of deliberate practice (Ericsson et al., 1993; Gladwell, 2008) or ten years (Hayes, 1989; H. A. Simon & Chase, 1973). While many researchers disagree with both of these figures, they all agree that there is still a large amount of painstaking effort and time spent in order to develop expertise (Baker & Young, 2014; Hambrick & Hoffman, 2016; Hambrick et al., 2014). In fact, a distinguishing characteristic of an expert is their ability to consistently and reliably display superior performance upon demand (Ericsson & Lehmann, 1996)

Weinstein (1993) argues for identification of two categories of expertise: (1) epistemic expertise, which is a function of what an expert knows, and (2) performative expertise, which is a function of what an expert does. Within the realm of cognitive science, expertise is defined in terms of development, knowledge structures, and reasoning processes (Hoffman, 1998). These loosely correlate with the categories identified by Weinstein in that development of expertise is gaining the skills necessary to perform a task, or performative expertise. Likewise, knowledge structures and their organization relate to epistemic expertise. However, Hoffman identified an additional attribute of expertise called "reasoning processes" that relates to the cognitive processes involved with making decisions.

Knowledge Organization

The organization of knowledge that an expert knows—their epistemic expertise—is much different than that of a novice. An expert's extensive knowledge affords them the ability to abstract both the knowledge and the problem, whereas novices have a much more simplistic grasp (Hmelo-Silver & Pfeffer, 2004). Furthermore, when discussing complex systems, experts tend to "have a more functional and behavioral understanding whereas novices, regardless of age, have a more structural representation" (Hmelo-Silver & Pfeffer, 2004, p. 132). Having greater knowledge, experts describe complex systems in terms of relationships, patterns, and outcomes, expressing a more integrative understanding (Chase & Simon, 1973a; Chi, Feltovich, & Glaser, 1981; D. P. Simon & Simon, 1978) affording experts the capability of distributing cognition into the context, whereby artifacts within the system are coupled together with the cognitive processes of the individual (Hutchins, 1995). Novices, on the other hand, describe these same complex systems more simplistically (Adelson, 1981, 1984), making reference to syntactically organized knowledge. For example, Hmelo-Silver and Pfeffer (2004) asked experts and children to create and then describe an aquarium. Experts discussed not only the physical structures that were present in their drawing but also their function, purpose, and behavior within the tank. The added dimensions of context and artifacts into the distributed cognitive processes further strengthens the important connections that couples epistemic knowledge with performative knowledge and reasoning.

Studies have shown that experts are also able to connect concepts and memories in meaningful ways. Their conceptual categories (Voss, Greene, Post, & Penner, 1983) are then able to be utilized to recognize patterns and provide the expert with intuition that conceptually different problem types may also exhibit the same characteristics (Murphy & Wright, 1984). For example, Groen and Patel (1988) found when comparing medical diagnosticians and medical students that experts tended to remember the essence of cases instead of their individual specifics.

Ericsson and Lehmann (1996) also found that when comparing experts to novices, experts are able to arrange problems into categories using features of their solutions, while novices are only able to utilize features of the problem statement alone. For example, Chi, Glaser, and Rees (1982) found that not only did experts possess more knowledge about physics than novices, but because of their superior knowledge structure organizations, experts could represent problems in terms of their relative principles. Conversely, novices were only able to express problems with regard to their surface elements.

The pattern recognition and interrelated problem categories possessed by experts affords them the ability to plan solutions, when possible, in memory and on-the-fly (VanLehn, 1996). Chase and Simon (1973b) demonstrated that an expert's pattern recognition accounted for the superior chess move selection and seemingly supernatural memory without breaking inherent human information processing limitations (e.g. limited short-term memory capacity) (Newell & Simon, 1972). Furthermore, Klein, Calderwood, and Clinton-Cirocco (1986) developed the Recognition Primed Decision (RPD) model that emphasizes the use of recognition for decision making rather than calculation or analysis. While studying fire ground commanders (FGCs), Klein et al. (1986) discovered that FGCs relied primarily on experience to identify an appropriate course of action rather than comparisons and evaluations of alternative options. FGCs were able to effectively identify when situations were typical (or not) and select the most effective action to take incorporating the distributed context into their cognitive process. Moreover, FGCs relied on their expertise to avoid meticulous internal deliberations when selecting an appropriate course of action (Klein, 1993).

Skill

Much of the relevant research with regards to performing as an expert measures expertise on a continuum from novice to expert (Abelson, 1981; Phelps & Shanteau, 1978). When engaging in deliberate practice, a novice first gains knowledge about the rules in order to be able to at least perform the skill being practiced. Through continued repetition and practice, the skill becomes almost innate. For instance, when learning to ride a bicycle, novices must learn how to balance, turn, pedal, and brake (i.e. the rules and functions of the bicycle). As a skill is practiced and developed there are "stage-like qualitative shifts that occur as expertise develops" (Hoffman, 1998, p. 84) whereby explicit instructions and knowledge become tacit or "automatic" (Sanderson, 1989).

When knowledge becomes tacit, it is more readily available for use in decision making and performing skill. There is a sort of muscle memory that develops that leads to decreased response times, quicker decision making, and more precise reactions (McLeod & Jenkins, 1991). These benefits are generally quantitatively measurable. For example, elite athletes when prompted with opportunities to perform can produce required reactions faster and earlier in the process than athletes with less skill. The reasoning for this difference is that tacit knowledge provides experts with the ability to perform anticipatory movements (Helsen & Pauwels, 1993) within a much smaller reaction window, often with greater accuracy and precision (McLeod & Jenkins, 1991).

In an analysis of expert jugglers, Huys, Daffertshofer, and Beek (2004) demonstrated that with a partially obstructed view and thus shorter available reaction times, experts were able to make necessary movement corrections while only able to view an object's apex. Furthermore, the researchers found that an expert juggler has a much lower variability to object trajectory patterns minimizing the corrections needed to keep the objects aloft in the first place. Likewise, in studying typing experts it was shown that experts tended to read well ahead of the text that they are typing (Gentner, 1988) and were able to link together prepared movements of their fingers in advance (Gentner, 1983; Salthouse, 1984), which led to an increase in typing speed as compared to those of novice typists. In fact, Salthouse (1984) determined that when an expert is disallowed the ability to look ahead, their typing skill was nearly reduced to that of a novice. Furthermore, Klein (1993) concluded that an expert's ability to mentally simulate a situation affords them to the capability to discover the optimal solution to problems quickly rather than having to compare several options. As a result, experts often identify a reasonable solution quickly, when considering what is known, and therefore do not necessarily need to generate alternative course of actions.

To further illustrate the look ahead capability of experts, researchers have found that expert sight-readers—those that can play an unfamiliar piece of music on an instrument given the musical score—tend to look further ahead to anticipate music notes and movements (Bean, 1938; Goolsby, 1994) similar to expert typists. Additionally, novice sight-readers often look at single individual notes, whereas experts tend to look at chunks of notes throughout the score of music. Consequently, experts are able to utilize their advanced knowledge of music theory, which "facilitates the efficiency of encoding of patterns and chunks" in anticipation of upcoming musical notes (Lehmann & Ericsson, 1996, p. 5).

Cognitive Reasoning

Similar to the phases of skill acquisition, cognitive reasoning in an individual progresses in stages (Kim, Ritter, & Koubek, 2013). In the early stages of learning, the novice spends their time attempting to grasp knowledge about the domain without applying it. This phase of development is characterized by reading, studying, and other methods of acquiring information. As practice continues, individuals begin to transition from knowledge acquisition to application whereby their attention shifts toward solving problems (VanLehn, 1996).

The study of expert performance began with the study of expert chess players and has continued to be the focus domain for decades (see Chase & Simon, 1973a; De Groot, 1965). In his seminal work, De Groot (1965) examined the cognitive processes of expert and novice chess players using think aloud strategies as they selected the best move when presented with various chess boards. Results showed that while both world-class chess players and those with considerably less skill performed planning and cognitive searches, the experts consistently selected the best move because the chess board is internally assimilated into the expert's cognitive processing. Conversely, those with less skill often failed to consider the best move despite undergoing the same cognitive processes. Further study (Chase &

Simon, 1973b) of the phenomenon determined that expert chess players did not generate the best selection during cognitive searching but rather by memory recall. Visual cuing by the chess board prompted the experts—those with considerably more practice and exposure to chess move permutations—to simply recall the best move from a previous example.

Anderson (1993) claims the ability to "speed up" is due to knowledge being converted from declarative knowledge (i.e. what they know) into procedural knowledge (i.e. what they can do), and "the speed of the individual pieces of procedural knowledge also increase with practice" (VanLehn, 1996, p. 25). Other researchers (Newell, 1994; Newell & Rosenbloom, 1981) have shown that this process of practice and experience gradually allows an individual to integrate several smaller pieces of knowledge into larger subsystems for the specific tasks to be accomplished. The larger subsystems of knowledge can then be applied to a problem or task with the addition of only a few pieces of declarative knowledge, which increase efficiency and speed.

Expertise Distributed in Context

In the formative years of research on expertise, chess was the dominant domain of interest (see De Groot, 1965, 1966). Chess is a highly organized and rules-based context that is ideal for research but lacks the complexity of actual human activity. Since all human activity takes place within a complex environment (Feltovich, Ford, & Hoffman, 1997), chess simplified this environment, facilitating the study of expertise. Many of the findings within the chess context were important and supported research into more complex environments (Lewandowsky & Kirsner, 2000).

Context has been shown to be an important element to consider when studying knowledge and expertise resulting from its role in codifying and structuring knowledge within an individual. Perception of the environment provides cues that prompt knowledge recall and application. Nassehi (2004) articulates that there is a gap between knowing and doing (application). Novices that might possess knowledge or skill but do not act or are not able to accurately apply and use it present a real problem. Moreover, just because someone possesses a particular piece of knowledge does not mean that it gets transferred, recalled, or applied when appropriate.

Bransford, Franks, Yve, and Sherwood (1989) define this possessed but not accessed knowledge as "inert knowledge" in that the knowledge "is accessed only in a restricted set of contexts even though it is applicable to a wide variety of domains" (Bransford et al., 1989, p. 472). For example, Carraher, Carraher, and Schliemann (1985) found that street vendors in Chile accustomed to mentally calculating a customer's bill were unable to perform the same calculations when removed from the context without the visual cues of handling the product. Unfamiliar with the usefulness of the calculations the vendors daily employed, they were unable to apply them to the tests given by the researchers. The disconnect between context and cognitive reasoning can be addressed during the learning process via examples that serve to give deeper meaning to the knowledge being acquired in order to foster its use when faced with similar problems (van Gog, Paas, & van Merrienboer, 2004; Kaminski, Sloutsky, & Heckler, 2008). Specifically, examples reinforce the "why" and the "how" information is (and can be) used, providing grounding in the various uses of the knowledge, as well as, the rationale for each step in the cognitive reasoning process (van Gog et al., 2004). Experts, having been exposed to several scenarios and examples, are able to consistently and effectively access and thereby apply their knowledge both in familiar and in unfamiliar situations.

Furthermore, Choi and Hannafin (1995, p. 53) state that knowledge is a result of "unique relationships between an individual and the environment" implying that the environment is an integral part of an individual's cognitive process. Removal from the context likewise removes the perceptual cues that are provided by the environment and the requisite interactions between the context and the team (Evans & Garling, 1991; Kaplan, 1991). Context is therefore important in "establishing meaningful linkages with experience and in promoting connections among knowledge, skill, and experience" (Choi & Hannafin, 1995, p. 54). Likewise, learning that occurs without context is less likely to be accessed and applied in unrelated situations (Black, Segal, Vitale, & Fadjo, 2012; Carraher et al., 1985). Consequently, learning that occurs through the use of contextually relevant problems and authentic tasks promotes engagement, motivation, and learning in students (Choi & Hannafin, 1995; Dochy, Segers, Van den Bossche, & Gijbels, 2003; Wilson, 1993).

Since human activity is situated within a given context, the environment can allow an individual to off-load some of their cognitive tasks (i.e. through the use of alarms, signs, and other environmental cues), which is important in complex work domains (Hollan et al., 2000; Smith & Collins, 2010). Responding to an alarm system or system notifications within a complex work environment, for example, is situated within the specific context and decisions are made internally but in concert with the external inputs presented to the individual (Shattuck & Miller, 2004, 2006).

Expertise Distributed in Teams

Just as context is an important aspect of expertise to consider, so too are tasks that involve teams of people working together. In organizations or tasks that utilize teamwork, effectively managing and coordinating the expertise distributed among the team members is important to ensure quality output (Faraj & Sproull, 2000). Not only does teamwork require expertise to perform tasks, but there must also be an awareness of where expertise is located within the team itself, where expertise is needed, and how it is to be applied to the task at hand. Faraj and Sproull (2000) found that coordination of expertise was strongly related with team performance (i.e. teams that perform well also optimally coordinate their expertise).

In team-oriented complex fields of practice requiring peak performance, much research has been conducted on team performance elucidating the fact that teams having experience working together generally perform better than those having no prior experience with their teammates (Faraj & Sproull, 2000; Larson, Christensen, Abbott, & Franz, 1996; Lewis, 2003; Moreland, Argote, & Krishnan, 1996; Smith-Jentsch, Kraiger, Cannon-Bowers, & Salas, 2009). Much of this success comes from an accurate, shared understanding of the context, tasks, and problems (Mathieu, Heffner, Goodwin, Salas, & Cannon-Bowers, 2000), awareness of the common knowledge overlap, generally known as a shared mental model (SMM)

(Cannon-Bowers, Salas, & Converse, 1993), and knowledge of where information can be elicited from within the team (Reagans, Argote, & Brooks, 2005).

As a result, the expertise gained through prolonged cooperation among team members has been shown to increase levels of trust (Uzzi, 1996), which in turn leads to a greater degree of information sharing (Uzzi & Lancaster, 2003). Consequently, increased trust allows an alignment of mental models and fosters an awareness of the knowledge and expertise availability distributed throughout the team.

Research has also shown that the degree of information sharing, specifically the sharing of "private" information, affects the decision-making processes within teams (Bowman & Wittenbaum, 2012; Lu, Yuan, & McLeod, 2012). When information is shared fluidly between team members, it has been observed that teams tend to make decisions more quickly and accurately due to everyone having a common mental picture, or team mental model, of the problem, context, and possible solutions (Mathieu et al., 2000; Mohammed & Dumville, 2001). As individuals work together in a team, they develop a conceptual picture of the skills and knowledge possessed by the other members on the team (Cannon-Bowers & Salas, 2001; Fiore, Salas, Cannon-Bowers, & London, 2001; Mohammed & Dumville, 2001). As a result, individuals are aware of who is best suited to provide a specific piece of expertise or skill to be leveraged in response to a problem, task, or event.

Moreover, research has shown that cross-training employees has a desirable impact on team performance as a result of individuals being trained in the skills and responsibilities of each other's roles within the organization (i.e. team expertise) (Cannon-Bowers, Salas, Blickensderfer, & Bowers, 1998; Marks, Sabella, Burke, & Zaccaro, 2002; Volpe, Cannon-Bowers, Salas, & Spector, 1996). Entin and Serfaty (1999) acknowledge that cross-training provides a framework for increasing the accuracy of SMMs, providing each individual with an explicit view of their teammates' roles, responsibilities, and expertise. However, they also recognize that, while cross-training positively influences SMMs, for the most benefit to team performance training must afford teams an opportunity "to exercise their SMMs through specific training of coordination strategies" (Entin & Serfaty, 1999, p. 324). In other words, teams must be trained together in situations that replicate their working environment, tasks, and necessary intra-team interactions (i.e. coordination strategies) to ensure that the training is authentic. Cross-training, therefore, not only positively impacts task specific expertise, but also fosters team-oriented expertise.

MEASURING AND CAPTURING EXPERTISE

One of the difficult endeavors when studying expertise is its measurement, especially when considering the domain specificity associated with being an expert. Also, while expertise may be available, it may only have a limited application and may not translate well to similar domains that have conditions that exceed the expertise a person has. Furthermore, the validity of such measures could potentially be affected by cognitive biases, prejudices, emotion, and cultural factors. Therefore, expertise has boundary conditions and its generalizability may be coupled to its degree of bandwidth. If expertise is measured using a narrow spectrum only then it may not generalize well to other fields of practice. As a result, there are many measurements that have been created that measure expertise, but they are all domain specific and have no generalizability outside of the target domain.

Quantitative Measures

Despite the inherit challenges to measuring expertise, attempting to do so provides many tangible benefits. When expertise is quantifiable, it can be used to identify experts among a pool of individuals within a given subject (Boeva, Krusheva, & Tsiporkova, 2012) or to compare the expertise of one individual to another (e.g. hiring new employees) (Royer, Carlo, Dufresne, & Mestre, 1996; Shanteau, Weiss, Thomas, & Pounds, 2002). Quantitative measurement is often trivial in domains and activities that have a visible ground truth with "fixed capabilities and limited moves, [and] the goal is unambiguous" (Serfaty, MacMillan, Entin, & Entin, 1997, p. 233), such as the formative research on expertise among chess players. For problems that are ill-defined, complex, ambiguous, and reliant on information availability, methodologies have been developed that seek to bridge the gap between the observational data inherent in qualitative measurement the difficulty of reliably measuring expertise.

CWS Performance Index

One such quantitative measurement index developed to measure expertise, the Cochran-Weiss-Shanteau (CWS) Performance Index developed by Cochran, Weiss, and Shanteau (Cochran, 1943; Weiss & Shanteau, 2003), has been used to assess expertise based on the premise that "expert judgment involves discrimination—seeing fine graduations among the stimuli and consistency evaluating similar stimuli similarly" (Germain & Tejeda, 2012, p. 206). The general idea of the CWS is that experts must be able to discriminate consistently. Measuring an individual's response to different stimuli can gauge discrimination while response variances measure the inconsistency. When combined this ratio (i.e. discrimination to consistency) gives a quantifiable measurement of performance expertise. Moreover, the CWS Performance Index can be used to measure expertise in teams (a single measurement for the entire team) and individuals (a single-subject score). Indeed, it has been used to measure the expert performance of air traffic controllers (Thomas, Willem, Shanteau, Raacke, & Friel, 2001), ergonomists, agricultural judges, and occupational therapists.

However, while the CWS Performance Index has been successfully applied within various contexts, it has limitations. First, CWS is interpreted relative to other experts and is not interpreted absolutely. As such, CWS can be used to compare the expertise of individuals to determine which is performing better (Germain & Tejeda, 2012) but only relative to each other. This presents an issue when even experts perform consistently incorrectly to stimuli. While their CWS Performance Index score might be high, indicative of an "expert," in actuality they may not be acting appropriately. Therefore, an expert might receive a high CWS Performance Index score, but that does not guarantee expertise. Germain and Tejeda explain that "a dance judge who evaluates the contenders primarily on the basis of appearance, taking into account hairdo and outfit very heavily, would be deemed an expert according to the CWS

Performance Index if those attributes were used to discriminate consistently among the dancers" (Germain & Tejeda, 2012, p. 207). In other words, the judge would consistently discriminate using the same criteria for each dancer, yielding a high discrimination/inconsistency ratio despite not accounting for skill of the dancer. However, it is obvious that this is not a true measure of the skill of the dancer and therefore we would not be able to say that we are measuring true expertise.

Inference Verification Technique

The Inference Verification Technique (IVT) developed by Royer, Carlo, Dufresne, and Mestre (1996) is another quantitative measure which can be used to discriminate between experts and novices. To measure expertise in a particular domain, an individual is given a corpus of text and asked to make inferences from two items taken from the text and creating an inference that connects the two items, called near inferences. Secondly, the individuals are then asked to connect an item of text with prior knowledge about the domain of interest and drawing valid inferences between the two items, called *far inferences*. Royer et al. (1996) found that experts were able to see underlying principles to problem solutions, but novices were only able to see the surface problem elements.

Thurstonian Model

The Thurstonian Model, developed by Steyvers, Miller, Lee, and Hemmer (2009), relies on the crowd as applied to order data, in that the average answers to ordering of data by a crowd is at least as good or better than that of each of the individual answers. When developing the model, it is important to weight the answers of those with more expertise and experience higher than others. However, the model is designed to account for individual differences as the distributions of answers are allowed to vary, which accommodates differences between the individuals while still capturing information about the objective ground truth from the crowd.

The Thurstonian Model has been shown to reliably identify the expert among a large pool of participants and can reliably measure expertise that "correlates highly with the actual accuracy of the answers" (Lee, Steyvers, de Young, & Miller, 2012, p. 3). However, this means that the model requires a large pool of participants from which to draw answers. Additionally, when ground truth might not necessarily be known, the model can be applied to represent ground truth as drawn from the crowd and can further be applied to prediction tasks. For example, the model could be used to ask individuals to predict end-of-season rankings for a sports team and then the information can be used to identify the expert ahead of time. Furthermore, the results could be used at the end of the year to further refine the model for subsequent seasons.

Qualitative Measures

Quantitative measurement of expertise is a worthwhile endeavor but has the potential to overlook some of the nuisances that occur within such a complex concept. Expertise is more than just performing to some gold standard. While an individual might possess extensive knowledge about a particular subject (i.e. they are a subject matter expert), they might not be able to perform tasks much better than a novice. Furthermore, in highly dynamic and complex environments that are uncertain, multidimensional, and (potentially) dangerous, quantitative measurement is difficult or near impossible. Dynamic environments stand starkly contrasted from more stable environments, such as the early research into the expertise of chess players (De Groot, 1966) as compared to the activities of operational planners (Rasmussen, Sieck, & Smart, 2009). Fortunately, dynamic environments can still provide useful knowledge with regards to expertise through qualitative measurement. Qualitative measurement can be employed in order to gather an understanding of the tasks, context, and knowledge necessary to perform as an expert. In contrast to traditional research on expertise, researchers sought to study how experts make decisions in their natural contexts or in simulations that captured the essential elements of their environment (Zsambok, 1997). To that end, naturalistic decision making (NDM) research studies "the way people use their experience to make decisions in field settings" (Zsambok, 1997, p. 4). Many qualitative models for measuring and studying expertise developed within the NDM research domain (e.g. mental models, recognition/metacognition model, recognition-primed decision model).

Mental Models

Used in several disciplines for several years, mental models are the organized knowledge structures within individuals that afford the capability to describe and make predictions about the physical environment with which they are interacting (Greca & Moreira, 2000). Mental models also enable individuals to recognize and remember relationships among the various components of their surrounding environment (Mathieu et al., 2000). Furthermore, mental models provide a vehicle for experts to perform mental simulation, thereby making effective and accurate decisions, even when under time stress. In fact, Cannon-Bowers and Salas (2001) found that in time-sensitive and high-stress environments that present time constraints that hinder planning, accurate mental models are crucial for optimal performance. Furthermore, not only do individuals have mental models that can be expressed, so too do teams of individuals. These intra-team mental models, or shared mental models, allow teams to adapt quickly to rapidly changing environments while still performing their job effectively (Cannon-Bowers et al., 1993).

Capturing and measuring mental models can be accomplished by a variety of qualitative methodologies depending on the phenomenon being studied, such as field observations (Button & Sharrock, 2009), think aloud protocols (Halasz & Moran, 1983), and cognitive mapping, to name a few (Kolkman, Kok, & van der Veen, 2005). Often, in order to fully capture a complex environment, it may be necessary to utilize multiple methodologies (Kraiger & Wenzel, 1997) and the researcher must be prepared to justify the choice of technique used as directed by their research questions (Mohammed, Klimoski, & Rentsch, 2000). These qualitative methodologies provide an opportunity to capture not only the knowledge and expertise required to perform complex tasks in dynamic environments, but can examine the interactions and importance of the context to the overall work.

Early research on shared mental models used the results to retrospectively explain performance differences among teams working on related tasks (Kleinman & Serfaty, 1989). However, more contemporary studies have been conducted in an effort to measure expertise and knowledge more directly. Results have shown that "teamwork and taskwork related positively to team process and performance" (Mohammed & Dumville, 2001, p. 91).

Recognition-Primed Decision Model

The recognition-primed decision (RPD) model was developed through study of firefighters to understand how these experts made decisions when dealing with time pressure and uncertain situations. Research for the RPD was conducted by employing probing question-based interviews with firefighters that had an average experience of 23 years (Klein, Calderwood, & Macgregor, 1989). Results found that participants overwhelmingly selected the first course of action that they identified rather than comparing a multiplicity of options that may be available. The tendency in critical situations where time pressure or uncertainty are involved is to "go with what we know" based on our prior experience. In this way, expertise provides a framework for mitigating problems rather than retrieving an analog, although that is not to say that analogical reasoning is not occurring (Lipshitz, Klein, Orasanu, & Salas, 2001).

The RPD model suggests that given time pressure and uncertainty or ill-defined goals, experts are able to work forward from existing situations, rather than working backward from a goal state to the current situation. Working forward allows the expert to continually adapt their course of action with regard to the current state as patterns are recognized (Lipshitz et al., 2001). Conversely, novices and intermediate individuals tend to work backward from a desired goal state to the current situation. Unfortunately, this approach falls apart in dynamic situations as the process must be restarted multiple times to work backward from the desired goal to the now-existing circumstance.

Recognition/Metacognition Model

The RPD model relies heavily on an expert's ability to recognize patterns and intuitively make decisions in spite of time pressure and uncertainty. However, when recognition is insufficient in the absence of discernable patterns for the current situation, the RPD falls short. Consequently, the recognition/metacognition (R/M) model, an extension the RPD model, adds an additional component that addresses the cognitive processes that occur in these unrecognizable situations (Cohen, Adelman, Tolcott, Bresnick, & Freeman, 1994). Cohen et al. (1994) suggest that when experts are presented with situations that are unrelated to prior patterns, cognitively they engage in metarecognition tasks, whereby they critique their understanding of the problem, correct their mental models, and reassess their course of action considering time available, cost of any errors, and their degree of uncertainty.

The R/M model provides a framework for understanding how expert decision makers test and improve the results of their pattern recognition and solution application (Cohen, Freeman, & Thompson, 1997). According to the R/M model, experts handle complex and dynamic situations and work within these unfamiliar situations by engaging in metacognitive strategies as they work forward towards an effective course of action. While this process is occurring, expert decision makers will continue to evaluate their current knowledge and expertise identifying knowledge gaps and are cognizant of the dangers of excessive trial and error.

DISCUSSION

While many researchers disagree with the amount of effort and time required to become an expert, they all agree that it takes considerable time and effort to do so. Furthermore, expertise is gained through deliberate and consistent practice. Deliberate practice prepares an individual to understand the rules and functions necessary to perform a particular task, develop organized knowledge structures, and promotes cognitive reasoning skills. Furthermore, increased experience gained through practice allows an expert to perform a task consistently, reliably, and accurately. Experts are able to apply these abilities in complex situations and are able to make connections to prior knowledge allowing knowledge to be transferred.

Despite its importance to society as a whole, expertise is a difficult concept to measure in any generalized manner due to the domain specificity of the concept. Many of the measures do a fair job at measuring the construct within their particular research domain but lack overall generalizability to the larger population. However, that does not mean that we should not continue to use these measures as they provide valuable empirical evidence of expertise despite specificity. The data can be used to quantifiably identify experts, which could be beneficial for identifying colleagues to collaborate with on interdisciplinary research topics or make predictive analyses of possible experts based on their responses to specialized domain-specific problem sets.

Despite the difficulty in quantifying expertise, naturalistic decision models can provide researchers a qualitative framework for analyzing complex and dynamic contexts and tasks. Complex tasks can be observed in their natural contexts so that not only are the tasks explored, so too are the interactions of individuals with each and the context directly. Other qualitative processes, such as think aloud protocols, can provide the researcher a view of the decision-making process of an expert while it occurs, rather than relying on respective recollection of the task. Therefore, real-time data can be gathered and the cognitive processes of the expert can be further explored by the researcher as it occurs. Quantitative measures typically lack this explorative nature and therefore are not flexible enough to capture the intricacies of complex environments.

FUTURE RESEARCH AGENDA

One of the fundamental problems surrounding the research of experts are the practicalities and logistics involved in such endeavors. When performing research within a specific domain, recruiting expert participants can be costly and inefficient. Recruitment becomes increasingly more difficult if a study is to be replicated across multiple domains. In an effort to alleviate these recruitment challenges and to employ comparisons between novices and experts, we present a framework for

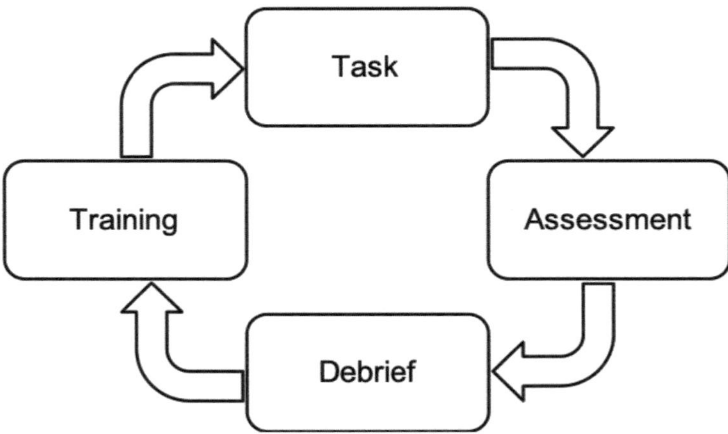

FIGURE 8.1 Creation of "convenient experts."

the comparison of novices and experts within a variety of domains. To address the difficulty of recruiting experts and then comparing the results with a novice, we propose the creation of "convenient experts" (see Figure 8.1). Typically, in academic research students are recruited as participants as a result of convenience due to their (1) quick and easy availability and (2) cost effectiveness. In the early stages of research, this has been shown to be sufficient to at least explore concepts at a cursory level. However, since expertise is domain specific, finding students with expertise in a particular domain (e.g. the chemical processing industry) is problematic at best. Hence, the problem that exists is that the results of many experiments/studies are predicated on novice—not expert—performance. What would be useful is to provide expert performance in studies so as to compare with novice performance in meaningful ways.

Studies have shown that real-world simulations modeled after a context have proven to provide a rich research environment for performing empirical research within a controlled laboratory setting. Furthermore, simulations provide flexibility in testing scenarios, hypotheses, and policies. Additionally, simulations can be used as an educational platform for training (Gallagher et al., 2005; Johansson, Trnka, Granlund, & Götmar, 2010), decision-making tool (Sheridan & Parasuraman, 2005), and as a planning model (Paul, Reddy, & DeFlitch, 2010). For example, the Living Lab Framework (LLF) developed by McNeese and colleagues (McNeese, Mancuso, McNeese, Endsley, & Forster, 2013) has successfully been deployed in multiple research studies over the last 15 years to further a variety of research agendas. The LLF methodology is an ecological psychology approach that presents a flexible framework useful for designing and evaluating technology in an authentic simulated scaled-world environment. It is useful for gaining a deep understanding of real-world contexts and identifying problems, creating solutions relevant solutions, and then evaluating them. For instance, the framework has been utilized in several studies to analyze such varying domains as crisis management (Hamilton et al., 2010; Jones,

2007; McNeese et al., 2005), military command and control (Hellar & Hall, 2009; Jones, McNeese, Connors, Jefferson, & Hall, 2004), and cybersecurity (Mancuso, Minotra, Giacobe, McNeese, & Tyworth, 2012; McNeese et al., 2013; Minotra & McNeese, 2017).

In addition to being deployed across a multitude of domains, it has also been instrumental in conducting research on varying theoretical perspectives using a plethora of methodologies. For example, team mental models and situation awareness have been studied using intelligent group interfaces and fuzzy cognitive maps. Other theories explored include transactive memory, task prioritization, information overload, and workload using virtual storytelling, geo-collaborative interfaces, shared workspaces, and cyber-visualizations from (McNeese et al., 2013)

To alleviate the difficulties associated with recruiting experts, we propose a process whereby we create domain-specific experts utilizing students. Achieving creation of experts will be accomplished through a regimented training program that takes advantage of real-world simulations of the target research domain[2] modeled after a particular research domain. The training will consist of providing the students with foundational information needed in order to perform tasks normally reserved for experts. Following the completion of the declarative knowledge training, several simulations will be given to familiarize participants with the tasks and how they are performed (i.e. procedural).

For example, NeoCITIES, a scaled-world environment developed the MINDS Group at Penn State University, could be used for the simulation portion of the research. NeoCITIES has enjoyed a lengthy career as a testbed for research in many domains covering numerous theoretical perspectives. NeoCITIES underlying architecture provides an adaptive framework giving the researcher broad latitude in modifying the user interface to support their particular research interests.

NeoCITIES has the capability to calculate team-based and individual performance measures. These performance measures can be utilized to identify experts once training and practice no longer provide measurable benefit and progression to performance. Once experts have been identified[3] they can then be used for comparative studies alongside novices that receive limited training and exposure. Being able to run comparative analyses on novices and experts provides a basis for testing interface design changes and its effects on performance for both experts and novices.

CONCLUSION

The demographics of the modern workplace are changing as the Baby Boomer generation nears retirement (Kuyken, Ebrahimi, & Saives, 2009; Spitulnik, 2006; Vu, 2006). In environments increasingly supported by complex socio-technical systems and characterized by interdependent workflow processes, ensuring the continuity of employee skills and knowledge is critical (Hinds et al., 2001; Vashisth, Kumar, & Chandra, 2010). If left unattended, this gap in specialized knowledge could lead to the loss of human capital, talent shortages, and increased costs associated with training younger talent. In a 2015 Center for Energy Workforce Development (CEWD) survey,[4] over 35% of the workforce was age 53 and above. As such, organizations should be preparing a means of capturing the already-possessed skills and knowledge of its aging workforce (Davenport & Prusak, 2005; Spender & Grant, 1996). In industries that heavily rely on teams to monitor and maintain critical system infrastructure, such as in the chemical processing industry (e.g. natural gas and offshore drilling operations), military command and control, and aviation environments, it becomes even more imperative that steps be taken now to prevent loss of service and allow continued safe operation within these environments. The provided expertise framework can be utilized to examine this and other challenging complex and dynamic domains of practice in hopes of developing a deeper understanding of the challenges of learning and training within these domains.

However, within this chapter we provide a framework and methodology in order to create "convenient experts" comprised of students, which are convenient to recruit in higher education environments. First, we expose students to declarative knowledge training providing a foundational basis for performing domain-specific tasks. Utilizing specialized software, such as NeoCITIES, a simulated scaled-world environment, offers a context-specific environment for performing laboratory-controlled research, whereby participants are repeatedly exposed to the tasks (i.e. procedural knowledge). Their progress can then be tracked and used for analyses concerning their level of expertise. Upon establishing "experts," further study can be done comparing the "convenient experts" to novices regardless of the research domain of interest.

NOTES

1. While literature exists that describes expertise in a nonlinear fashion (see Araújo et al., 2010), in an effort to remain parsimonious this chapter will be using the more traditional novice/expert distinction unless noted otherwise as deemed necessary.
2. Given that the natural gas industry is plagued by an impending crisis of retirement and thereby a loss of expertise, we will be conducting research in this domain. It is our goal to understand this complex and dynamic environment so that knowledge acquisition and transfer can be facilitated through properly developed training protocols employing a real-world simulation of the environment that leverages current experts training alongside newly hired employees (i.e. novices), and further supplemented with an electronic cognitive aide.
3. One phase of the study will be to create "experts" through multiple exposures to the simulated environment and with training. The NeoCITIES scoring model will be used to identify the point that performance plateaus thereby indicating that a participant has reached the point of limited progression and are therefore now an expert (see Figure 8.1).
4. www.cewd.org/surveyreport/CEWD2015SurveySummary.pdf

REFERENCES

Abelson, R. P. (1981). Psychological status of the script concept. *American Psychologist*, *36*(7), 715–729. https://doi.org/10.1037/0003-066X.36.7.715

Adelson, B. (1981). Problem solving and the development of abstract categories in programming languages. *Memory & Cognition*, *9*(4), 422–433. https://doi.org/10.3758/BF03197568

Adelson, B. (1984). When novices surpass experts: The difficulty of a task may increase with expertise. *Journal of Experimental Psychology: Learning, Memory, and Cognition*, *10*(3), 483–495. https://doi.org/10.1037/0278-7393.10.3.483

Alexander, P. A. (2005). Teaching towards expertise. *British Journal of Educational Psychology*, *2*(3), 29–45.

Anderson, J. R. (1993). Rules of the mind. In *Rules of the mind* (pp. ix, 320 p.). New York, NY: Psychology Press. https://doi.org/10.4324/9781315806938

Araújo, D., Fonseca, C., Davids, K., Garganta, J., Volossovitch, A., Brandão, R., & Krebs, R. (2010). The role of ecological constraints on expertise development. *Talent Development and Excellence*, *2*(2), 165–179.

Baker, J., & Young, B. (2014). 20 years later: Deliberate practice and the development of expertise in sport. *International Review of Sport and Exercise Psychology*, *7*(1), 135–157. https://doi.org/10.1080/1750984X.2014.896024

Bean, K. L. (1938). An experimental approach to the reading of music. *Psychological Monographs*, *50*(6), i–80. https://doi.org/10.1037/h0093540

Bereiter, C., & Scardamalia, M. (1986). Educational relevance of the study of expertise. *Interchange*, *17*(2), 10–19.

Bishop, J. (1989). Occupational training in high school: When does it pay off? *Economics of Education Review*, *8*(1), 1–15. https://doi.org/10.1016/0272-7757(89)90031-9

Black, J. B., Segal, A., Vitale, J., & Fadjo, C. (2012). Embodied cognition and learning environment design. In D. Jonassen and S. Lamb (Eds.), *Theoretical foundations of student-centered learning environments* (Vol. 2). New York, NY: Routledge.

Boeva, V., Krusheva, M., & Tsiporkova, E. (2012). Measuring expertise similarity in expert networks. In *2012 6th IEEE International Conference Intelligent Systems* (pp. 53–57). Sofia, Bulgaria: IEEE. https://doi.org/10.1109/IS.2012.6335190

Bowman, J. M., & Wittenbaum, G. M. (2012). Time pressure affects process and performance in hidden-profile groups. *Small Group Research*, *43*(3), 295–314. https://doi.org/10.1177/1046496412440055

Bransford, J. D., Franks, J. J., Yve, N. J., & Sherwood, R. D. (1989). New approaches to instruction: Because wisdom can't be told. In S. Vosniadou & A. Ortony (Eds.), *Similarity and analogical reasoning* (pp. 470–497). Cambridge: Cambridge University Press. https://doi.org/10.1017/CBO9780511529863.022

Button, G., & Sharrock, W. (2009). Studies of work and the workplace in HCI: Concepts and techniques. In J. M. Carroll (Ed.), *Synthesis lectures on human-centered informatics* (Vol. 2, pp. 1–96). Williston, VT: Morgan & Claypool. https://doi.org/10.2200/S00177ED1V01Y200903HCI003

Cannon-Bowers, J. A., & Salas, E. (2001). Reflections on shared cognition. *Journal of Organizational Behavior*, *22*(2), 195–202. https://doi.org/10.1002/job.82

Cannon-Bowers, J. A., Salas, E., Blickensderfer, E., & Bowers, C. A. (1998). The impact of cross-training and workload on team functioning: A replication and extension of initial findings. *Human Factors*, *40*(1), 92–101. https://doi.org/10.1518/001872098779480550

Cannon-Bowers, J. A., Salas, E., & Converse, S. (1993). Shared mental models in expert team decision making. In *Individual and group decision making: Current issues* (Vol. 221, pp. 221–246). Hillsdale, NJ: Lea Lawrence Erlbaum Associates.

Carraher, T. N., Carraher, D. W., & Schliemann, A. D. (1985). Mathematics in the streets and in schools. *British Journal of Developmental Psychology*, *3*(1), 21–29. https://doi.org/10.1111/j.2044-835X.1985.tb00951.x

Chase, W. G., & Simon, H. A. (1973a). Perception in chess. *Cognitive Psychology*, *4*(1), 55–81. https://doi.org/10.1016/0010-0285(73)90004-2

Chase, W. G., & Simon, H. A. (1973b). The mind's eye in chess. In W. G. Chase (Ed.), *Visual information processing* (pp. 215–281). New York, NY: Academic Press. https://doi.org/10.1016/B978-0-12-170150-5.50011-1

Chi, M. T. H. H., Feltovich, P. J., & Glaser, R. (1981). Catagorization and representation of physics problems by experts and novices. *Cognitive Science*, *5*(2), 121–152. https://doi.org/10.1207/s15516709cog0502_2

Chi, M. T. H. H., Glaser, R., & Rees, E. (1982). Expertise in problem solving. In R. J. Sternberg (Ed.) *Advances in the psychology of human intelligence* (Vol. 1, pp. 7–75). Hillsdale, NJ: Erlbaum.

Choi, J.-I., & Hannafin, M. (1995). Situated cognition and learning environments: Roles, structures, and implications for design. *Source: Educational Technology Research and Development*, *43*(2), 53–69. https://doi.org/10.1007/Bf02300472

Cochran, W. G. (1943). The comparison of different scales of measurement for experimental results. *The Annal of Mathematical Statistics*, *14*(3), 206–216.

Cohen, M. S., Adelman, L., Tolcott, M. A., Bresnick, T. A., & Freeman, M. F. (1994). *A cognitive framework for battlefield commanders' situation assessment*. Arlington, VA: Cognitive Technologies Inc.

Cohen, M. S., Freeman, J. T., & Thompson, B. B. (1997). Training the naturalistic decision maker. In C. E. Zsambok & G. Klein (Eds.), *Naturalistic decision making* (pp. 257–268). Mahwah, NJ: Lawrence Erlbaum Associates, Inc.

Davenport, T. H., & Prusak, L. (2005). Working knowledge: How organizations manage what they know [Book Review]. *IEEE Engineering Management Review*, *31*(4), 301. https://doi.org/10.1109/EMR.2003.1267012

De Groot, A. D. (1965). *Thought and choice in chess*. The Hague: Mouton.

De Groot, A. D. (1966). Perception and memory versus thought: Some old ideas and recent findings. In B. Kleinmuntz (Ed.) *Problem solving: Research, method and theory* (pp. 19–51). New York, NY: John Wiley.

Dochy, F., Segers, M., Van den Bossche, P., & Gijbels, D. (2003). Effects of problem-based learning: A meta-analysis. *Learning and Instruction*, *13*(5), 533–568. https://doi.org/10.1016/S0959-4752(02)00025-7

Entin, E. E., & Serfaty, D. (1999). Adaptive team coordination. *Human Factors*, *41*(2), 312–325. https://doi.org/10.1518/001872099779591196

Ericsson, K. A., Krampe, R. T., & Tesch-Römer, C. (1993). The role of deliberate practice in the acquisition of expert performance. *Psychological Review*, *100*(3), 363–406. https://doi.org/10.1037/0033-295X.100.3.363

Ericsson, K. A., & Lehmann, A. C. (1996). Expert and exceptional performance. *Annual Review of Psychology*, *47*(1), 273–305.

Ericsson, K. A., Prietula, M. J., & Cokely, E. T. (2007). The making of an expert. *Harvard Business Review*, 1–9.

Evans, G. W., & Garling, T. (1991). Environment, cognition, and action: The need for integration. In T. Garling & G. W. Evans (Eds.), *Environment, cognition, and action: An integrated approach* (pp. 3–14). New York: Oxford University Press.

Faraj, S., & Sproull, L. (2000). Coordinating expertise in software development teams. *Management Science*, *46*(12), 1554–1568. https://doi.org/10.1287/mnsc.46.12.1554.12072

Feltovich, P. J., Ford, K. M., & Hoffman, R. R. (1997). Expertise in context. *Menlo Park*, 590.

Fiore, S. M., Salas, E., Cannon-Bowers, J. A., & London, M. (2001). Group dynamics and shared mental model development. In *How people evaluate others in organizations.* (pp. 309–336). Mahwah, NJ: Lawrence Erlbaum Associates.

Fischer, A., Greiff, S., & Funke, J. (2012). The process of solving complex problems. *Journal of Problem Solving*, *4*(1), 19–42. https://doi.org/10.7771/1932-6246.1118

Gallagher, A. G., Ritter, E. M., Champion, H., Higgins, G., Fried, M. P., Moses, G., . . . Satava, R. M. (2005). Virtual reality simulation for the operating room. *Annals of Surgery*, *241*(2), 364–372. https://doi.org/10.1097/01.sla.0000151982.85062.80

Gentner, D. R. (1983). The acquisition of typewriting skill. *Acta Psychologica*, *54*(1–3), 233–248. https://doi.org/10.1016/0001-6918(83)90037-9

Gentner, D. R. (1988). Expertise in typewriting. In *The nature of expertise* (pp. 1–21). Mahwah, NJ: Lawrence Erlbaum Associates.

Germain, M.-L., & Tejeda, M. J. (2012). A preliminary exploration on the measurement of expertise: An initial development of a psychometric scale. *Human Resource Development Quarterly*, *23*(2), 203–232. https://doi.org/10.1002/hrdq.21134

Gick, M. L. (1986). Problem-solving strategies. *Educational Psychologist*, *21*(1–2), 99–120. https://doi.org/10.1207/s15326985ep2101&2_6

Gladwell, M. (2008). *Outliers: The story of success*. New York: Little, Brown and Company.

Goldman, S. R., & Petrosino, A. J. (1999). Design principles for instruction in content domains: Lessons from research on expertise and learning. In F. T. Durso (Ed.), *Handbook of applied cognition* (pp. 595–628). Chichester: Wiley.

Goolsby, T. W. (1994). Eye movement in music reading: Effects of reading ability, notational complexity, and encounters. *Music Perception: An Interdisciplinary Journal*, *12*(1), 77–96. https://doi.org/10.2307/40285756

Greca, I. M., & Moreira, M. A. (2000). Mental models, conceptual models, and modelling. *International Journal of Science Education*, *22*(1), 1–11. https://doi.org/10.1080/095006900289976

Groen, G. J., & Patel, V. L. (1988). The relationship between comprehension and reasoning in medical expertise. In *The nature of expertise* (pp. 287–310). Mahwah, NJ: Lawrence Erlbaum Associates.

Halasz, F. G., & Moran, T. P. (1983). Mental models and problem solving in using a calculator. *Proceedings of the SIGCHI Conference on Human Factors in Computing Systems—CHI '83*, 212–216. https://doi.org/10.1145/800045.801613

Hambrick, D. Z., & Hoffman, R. R. (2016). Expertise: A second look. *IEEE Intelligent Systems*, *31*(4), 50–55. https://doi.org/10.1109/MIS.2016.69

Hambrick, D. Z., Oswald, F. L., Altmann, E. M., Meinz, E. J., Gobet, F., & Campitelli, G. (2014). Deliberate practice: Is that all it takes to become an expert? *Intelligence*, *45*, 34–45. https://doi.org/10.1016/j.intell.2013.04.001

Hamilton, K., Mancuso, V., Minotra, D., Hoult, R., Mohammed, S., Parr, A., . . . McNeese, M. (2010). Using the NeoCITIES 3.1 simulation to study and measure team cognition. *Proceedings of the Human Factors and Ergonomics Society Annual Meeting*, *54*(4), 433–437. https://doi.org/10.1177/154193121005400434

Hayes, J. R. (1989). Cognitive processes in creativity. *Handbook of Creativity*, *7*(18), 135–145.

Hellar, D. B., & Hall, D. L. (2009). NeoCITIES: An experimental test-bed for quantifying the effects of cognitive aids on team performance in C2 situations. *Proceedings of SPIE—The International Society for Optical Engineering*, *7348*. https://doi.org/10.1117/12.818797

Helsen, W., & Pauwels, J. M. (1993). The relationship between expertise and visual information processing in sport. *Advances in Psychology*, *102*(C), 109–134. https://doi.org/10.1016/S0166-4115(08)61468-5

Hesketh, A. J. (2000). Recruiting an elite? Employers' perceptions of graduate education and training. *Journal of Education and Work*, *13*(3), 245–271. https://doi.org/10.1080/713676992

Hinds, P. J., Patterson, M., & Pfeffer, J. (2001). Bothered by abstraction: The effect of expertise on knowledge transfer and subsequent novice performance. *Journal of Applied Psychology*, *86*(6), 1232–1243. https://doi.org/10.1037/0021-9010.86.6.1232

Hmelo-Silver, C. E., & Pfeffer, M. G. (2004). Comparing expert and novice understanding of a complex system from the perspective of structures, behaviors, and functions. *Cognitive Science*, *28*(1), 127–138. https://doi.org/10.1016/S0364-0213(03)00065-X

Hoffman, R. R. (1998). How can expertise be defined? Implications of research from cognitive psychology. In R. Williams, W. Faulkner, & J. Fleck (Eds.), *Exploring expertise* (pp. 81–100). London: Macmillan Press. https://doi.org/10.1007/978-1-349-13693-3_4

Hollan, J. J. D., Hutchins, E., & Kirsh, D. (2000). Distributed cognition: Toward a new foundation for human-computer interaction research. *ACM Transactions on Computer-Human Interaction*, *7*(2), 174–196. https://doi.org/10.1145/353485.353487

Hutchins, E. (1995). How a cockpit remembers its speeds. *Cognitive Science*, *19*(3), 265–288. https://doi.org/10.1016/0364-0213(95)90020-9

Huys, R., Daffertshofer, A., & Beek, P. J. (2004). Multiple time scales and multiform dynamics in learning to juggle. *Motor Control*, *8*, 188–212. https://doi.org/10.1123/mcj.8.2.188

Johansson, B. J. E., Trnka, J., Granlund, R., & Götmar, A. (2010). The effect of a geographical information system on performance and communication of a command and Control Organization. *International Journal of Human-Computer Interaction*. https://doi.org/10.1080/10447310903498981

Jones, R. E. T. (2007). The development of an emergency crisis management simulation to assess the impact a fuzzy cognitive map decision-aid has on team cognition and team decision-making. *Dissertation Abstracts International: Section B: The Sciences and Engineering*, *67*, 4521.

Jones, R. E. T., McNeese, M. D., Connors, E. S., Jefferson, T., & Hall, D. L. (2004). A Distributed cognition simulation involving homeland security and defense: The development of NeoCITIES. *Proceedings of the Human Factors and Ergonomics Society Annual Meeting*, *48*(3), 631–634. https://doi.org/10.1177/154193120404800376

Kaminski, J. A., Sloutsky, V. M., & Heckler, A. F. (2008). Learning theory: The Advantage of Abstract Examples in Learning Math. *Science (New York, N.Y.)*, *320*(5875), 454–455. https://doi.org/10.1126/science.1154659

Kaplan, R. (1991). Environmental description and prediction: A conceptual analysis. In T. Garling & G. W. Evans (Eds.), *Environment, cognition, and action: An integrated approach* (pp. 19–34). New York: Oxford University Press.

Kim, J. W., Ritter, F. E., & Koubek, R. J. (2013). An integrated theory for improved skill acquisition and retention in the three stages of learning. *Theoretical Issues in Ergonomics Science*, *14*(1), 22–37. https://doi.org/10.1080/1464536X.2011.573008

Klein, G. A. (1993). *A Recognition-Primed Decision (RPD) model of rapid decision making*. New York: Ablex Publishing Corporation.

Klein, G. A., Calderwood, R., & Clinton-Cirocco, A. (1986). Rapid decision making on the fire ground. *Proceedings of the Human Factors Society Annual Meeting*, *30*(6), 576–580. https://doi.org/10.1177/154193128603000616

Klein, G. A., Calderwood, R., & Macgregor, D. (1989). Critical decision method for eliciting knowledge. *IEEE Transactions on Systems, Man and Cybernetics*, *19*(3), 462–472. https://doi.org/10.1109/21.31053

Klein, G., & Borders, J. (2016). The ShadowBox approach to cognitive skills training: An empirical evaluation. *Journal of Cognitive Engineering and Decision Making*, *10*(3), 268–280. https://doi.org/10.1177/1555343416636515

Kleinman, D. L., & Serfaty, D. (1989). Team performance assessment decision making. In *Proceedings of the symposium on Interactive Networked Simulation for Training* (pp. 22–27). Orlando, FL: University of Central Florida.

Kolkman, M. J., Kok, M., & van der Veen, A. (2005). Mental model mapping as a new tool to analyse the use of information in decision-making in integrated water management.

Physics and Chemistry of the Earth, Parts A/B/C, 30(4–5), 317–332. https://doi.org/10.1016/j.pce.2005.01.002

Kraiger, K., & Wenzel, L. H. (1997). Conceptual development and empirical evaluation of measures of shared mental models as indicators of team effectiveness. *Team Performance Assessment and Measurement: Theory, Methods, and Applications*, 63–84.

Kuyken, K., Ebrahimi, M., & Saives, A.-L. (2009). Intergenerational knowledge transfer in high-technological companies: A comparative study between Germany and Quebec. In *Proceedings of the annual conference of the Administrative Sciences Association of Canada (ASAC), Niagara Falls, 6–9 June*. Montréal, Québec, Canada: Emerald.

Larson, J. R., Christensen, C., Abbott, A. S., & Franz, T. M. (1996). Diagnosing groups: Charting the flow of information in medical decision-making teams. *Journal of Personality and Social Psychology, 71*(2), 315–330. https://doi.org/10.1037/0022-3514.71.2.315

Lee, M. D., Steyvers, M., de Young, M., & Miller, B. (2012). Inferring expertise in knowledge and prediction ranking tasks. *Topics in Cognitive Science, 4*(1), 151–163. https://doi.org/10.1111/j.1756-8765.2011.01175.x

Lehmann, A. C., & Ericsson, K. A. (1996). Performance without preparation: Structure and acquisition of expert sight-reading and accompanying performance. *Psychomusicology: A Journal of Research in Music Cognition, 15*(1–2), 1–29. https://doi.org/10.1037/h0094082

Lewandowsky, S., & Kirsner, K. (2000). Knowledge partitioning: Context-dependent use of expertise. *Memory & Cognition, 28*(2), 295–305. https://doi.org/10.3758/BF03213807

Lewis, K. (2003). Measuring transactive memory systems in the field: Scale development and validation. *Journal of Applied Psychology, 88*(4), 587–604. https://doi.org/10.1037/0021-9010.88.4.587

Lipshitz, R., Klein, G., Orasanu, J., & Salas, E. (2001). Taking stock of naturalistic decision making. *Journal of Behavioral Decision Making, 14*(5), 331–352. https://doi.org/10.1002/bdm.381

Lu, L., Yuan, Y. C., & McLeod, P. L. (2012). Twenty-five years of hidden profiles in group decision making: A meta-analysis. *Personality and Social Psychology Review, 16*(1), 54–75. https://doi.org/10.1177/1088868311417243

Mancuso, V. F., Minotra, D., Giacobe, N., McNeese, M., & Tyworth, M. (2012). idsNETS: An experimental platform to study situation awareness for intrusion detection analysts. In *2012 IEEE International Multi-Disciplinary Conference on Cognitive Methods in Situation Awareness and Decision Support, CogSIMA 2012* (pp. 73–79). New Orleans, LA: IEEE. https://doi.org/10.1109/CogSIMA.2012.6188411

Marks, M. A., Sabella, M. J., Burke, C. S., & Zaccaro, S. J. (2002). The impact of cross-training on team effectiveness. *Journal of Applied Psychology, 87*(1), 3–13. https://doi.org/10.1037/0021-9010.87.1.3

Mathieu, J. E., Heffner, T. S., Goodwin, G. F., Salas, E., & Cannon-Bowers, J. A. (2000). The influence of shared mental models on team process and performance. *Journal of Applied Psychology, 85*(2), 273–283. https://doi.org/10.1037/0021-9010.85.2.273

McLeod, P., & Jenkins, S. (1991). Timing accuracy and decision time in high-speed ball games. *International Journal of Sport Psychology, 22*(3–4), 279–295.

McNeese, M. D., Bains, P., Brewer, I., Brown, C., Connors, E. S., Jefferson, T., & Terrell, I. (2005). The NeoCITIES simulation: Understanding the design and experimental methodology used to develop a team emergency management simulation. *Proceedings of the Human Factors and Ergonomics Society Annual Meeting, 49*(3), 591–594. https://doi.org/10.1177/154193120504900380

McNeese, M. D., Mancuso, V., McNeese, N., Endsley, T., & Forster, P. (2013). Using the living laboratory framework as a basis for understanding next-generation analyst work.

In B. D. Broome, D. L. Hall, & J. Llinas (Eds.), *SPIE 8758, next-generation analyst* (p. 87580F). Baltimore, MD: International Society for Optics and Photonics. https://doi.org/10.1117/12.2016514

Mehta, S., Suto, I., Elliott, G., & Rushton, N. (2011). Why study economics? Perspectives from 16–19-year-old students. *Citizenship, Social and Economics Education*, *10*(2–3), 199–212.

Minotra, D., & McNeese, M. D. (2017). Predictive aids can lead to sustained attention decrements in the detection of non-routine critical events in event monitoring. *Cognition, Technology & Work*, *19*(1), 161–177.

Mohammed, S., & Dumville, B. C. (2001). Team mental models in a team knowledge framework: expanding theory and measurement across disciplinary boundaries. *Journal of Organizational Behavior*, *22*(22), 89–106. https://doi.org/10.1002/job.86

Mohammed, S., Klimoski, R., & Rentsch, J. R. (2000). The measurement of team mental models: We have no shared schema. *Organizational Research Methods*, *3*(2), 123–165. https://doi.org/10.1177/109442810032001

Moreland, R. L., Argote, L., & Krishnan, R. (1996). Socially shared cognition at work: Transactive memory and group performance. In *What's social about social cognition? Research on socially shared cognition in small groups* (pp. 57–84).

Murphy, G. L., & Wright, J. C. (1984). Changes in conceptual structure with expertise: Differences between real-world experts and novices. *Journal of Experimental Psychology: Learning, Memory, and Cognition*, *10*(1), 144–155. https://doi.org/10.1037/0278-7393.10.1.144

Nassehi, A. (2004). What do we know about knowledge? An essay on the knowledge society. *The Canadian Journal of Sociology*, *29*(3), 439–449. https://doi.org/10.1353/cjs.2004.0043

Newell, A. (1994). *Unified theories of cognition*. Cambridge, MA: Harvard University Press.

Newell, A., & Rosenbloom, P. S. (1981). Mechanisms of skill acquisition and the law of practice. In J. Anderson (Ed.), *Cognitive skills and their acquisition* (pp. 1–56). Hillsdale, NJ: Lawrence Erlbaum Associates, Inc.

Newell, A., & Simon, H. A. (1972). *Human problem solving* (Vol. 104). Englewood Cliffs, NJ: Prentice-Hall.

Paul, S. A., Reddy, M. C., & DeFlitch, C. J. (2010). A systematic review of simulation studies investigating emergency department overcrowding. *Simulation*, *86*(8–9), 559–571. https://doi.org/10.1177/0037549710360912

Phelps, R. H., & Shanteau, J. (1978). Livestock judges: How much information can an expert use? *Organizational Behavior and Human Performance*, *21*(2), 209–219. https://doi.org/10.1016/0030-5073(78)90050-8

Rasmussen, L. J., Sieck, W. R., & Smart, P. R. (2009). What is a good plan? cultural variations in expert planners' concepts of plan quality. *Journal of Cognitive Engineering and Decision Making*, *3*(3), 228–252. https://doi.org/10.1518/155534309X474479

Reagans, R., Argote, L., & Brooks, D. (2005). Individual experience and experience working together: Predicting learning rates from knowing who knows what and knowing how to work together. *Management Science*, *51*(6), 869–881. https://doi.org/10.1287/mnsc.1050.0366

Royer, J. M., Carlo, M. S., Dufresne, R., & Mestre, J. (1996). The assessment of levels of domain expertise while reading. *Cognition and Instruction*, *14*(3), 373–408. https://doi.org/10.1207/s1532690xci1403_4

Salthouse, T. A. (1984). Effects of age and skill in typing. *Journal of Experimental Psychology: General*, *113*(3), 345–371. https://doi.org/10.1037/0096-3445.113.3.345

Sanderson, P. (1989). Verbalizable knowledge and skilled task performance: Association, dissociation, and mental models. *Journal of Experimental Psychology. Learning, Memory, and Cognition*, *15*(4), 729–747. https://doi.org/10.1037//0278-7393.15.4.729

Serfaty, D., MacMillan, J., Entin, E. E., & Entin, E. B. (1997). The decision-making expertise of battle commanders. In C. E. Zsambok & G. Klein (Eds.), *Naturalistic decision making* (pp. 233–246). New York, NY: Psychology Press.

Shanteau, J., Weiss, D. J., Thomas, R. P., & Pounds, J. C. (2002). Performance-based assessment of expertise: How to decide if someone is an expert or not. *European Journal of Operational Research*, *136*(2), 253–263. https://doi.org/10.1016/S0377-2217(01)00113-8

Shattuck, L. G., & Miller, N. L. (2004). A process tracing approach to the investigation of situated cognition. *Proceedings of the Human Factors and Ergonomics Society Annual Meeting*, *48*(3), 658–662. https://doi.org/10.1177/154193120404800382

Shattuck, L. G., & Miller, N. L. (2006). Extending naturalistic decision making to complex organizations: A dynamic model of situated cognition. *Organization Studies*, *27*(7), 989–1009. https://doi.org/10.1177/0170840606065706

Sheridan, T. B., & Parasuraman, R. (2005). Human-automation interaction. *Reviews of Human Factors and Ergonomics*, *1*(1), 89–129. https://doi.org/10.1518/155723405783703082

Simon, D. P., & Simon, H. A. (1978). Individual differences in solving physics problems. *Children's Thinking: What Develops*, 325–348.

Simon, H. A., & Chase, W. G. (1973). Skill in chess. *American Scientist*, *61*, 394–403. https://doi.org/10.1511/2011.89.106

Smith, E. R., & Collins, E. C. (2010). Situated cognition. In B. Mesquita, L. F. Barrett, & E. R. Smith (Eds.), *The mind in context* (pp. 126–148). New York: The Guildford Press.

Smith-Jentsch, K. A., Kraiger, K., Cannon-Bowers, J. A., & Salas, E. (2009). Do familiar teammates request and accept more backup? Transactive memory in air traffic control. *Human Factors*, *51*(2), 181–192. https://doi.org/10.1177/0018720809335367

Spender, J., & Grant, R. M. (1996). Knowledge of the firm: Overview. *Strategic Management Journal*, *17*(S2), 5–9. https://doi.org/10.1002/smj.4250171103

Spitulnik, J. J. (2006). Cognitive development needs and performance in an aging workforce. *Organization Development Journal*, *24*(3), 44–53.

Steyvers, M., Miller, B., Lee, M., & Hemmer, P. (2009). The wisdom of crowds in the recollection of order information. In *Twenty-third annual conference on Neural Information Processing System* (pp. 1785–1793) . Red Hook, NY: Curran Associates.

Thomas, R. P., Willem, B., Shanteau, J., Raacke, J., & Friel, B. (2001). CWS applied to controllers in a high fidelity simulation of ATC. In *International symposium on aviation psychology*. Columbus, OH: Ohio State University.

Tracey, J. B., Hinkin, T. R., Tran, T. L. B., Emigh, T., Kingra, M., Taylor, J., & Thorek, D. (2015). A field study of new employee training programs. *Cornell Hospitality Quarterly*, *56*(4), 345–354. https://doi.org/10.1177/1938965514554211

Urick, M. (2017). Adapting training to meet the preferred learning styles of different generations. *International Journal of Training and Development*, *21*(1), 53–59. https://doi.org/10.1111/ijtd.12093

Uzzi, B. (1996). The sources and consequences of embeddedness for the economic performance of organizations: The network effect. *American Sociological Review*, *61*(4), 674–698. https://doi.org/10.2307/2096399

Uzzi, B., & Lancaster, R. (2003). Relational embeddedness and learning: The case of bank loan managers and their clients. *Management Science*, *49*(4), 383–399. https://doi.org/10.1287/mnsc.49.4.383.14427

van Gog, T., Paas, F., & van Merrienboer, J. J. G. (2004). Process-oriented worked examples: Improving transfer performance through enhanced understanding. *Instructional Science*, *32*(411), 83–98. https://doi.org/10.1023/B:TRUC.0000021810.70784.b0

VanLehn, K. (1996). Cognitive skill acquisition. *Annual Review of Psychology*, *47*, 513–539. https://doi.org/10.1146/annurev.psych.47.1.513

Vashisth, R., Kumar, R., & Chandra, A. (2010). Barriers and facilitators to knowledge management: Evidence from selected indian universities. *The IUP Journal of Knowledge Management*, *VIII*(4), 7–27.

Volpe, C., Cannon-Bowers, J., Salas, E., & Spector, P. (1996). The impact of cross-training on team functioning: An empirical investigation. *Human Factors*, *1*, 87–100. https://doi.org/10.1037/0021-9010.87.1.3

Voss, J. F., Greene, T. R., Post, T. A., & Penner, B. C. (1983). Problem-solving skill in the social sciences. *Psychology of Learning and Motivation*, *17*, 165–213. https://doi.org/10.1016/S0079-7421(08)60099-7

Vu, Y. (2006). Unprepared for aging workers. *Canadian HR Reporter*, *19*(14).

Weinstein, B. D. (1993). What is an expert? *Theoretical Medicine*, *14*(1), 57–73. https://doi.org/10.1007/BF00993988

Weiss, D. J., & Shanteau, J. (2003). Empirical assessment of expertise. *Human Factors*, *45*(1), 104–116. https://doi.org/10.1518/hfes.45.1.104.27233

Wilson, A. L. (1993). The promise of situated cognition. *New Directions for Adult and Continuing Education*, *1993*(57), 71–79. https://doi.org/10.1002/ace.36719935709

Wright, G., & Bolger, F. (1992). *Expertise and decision support*. New York, NY: Plenum Press. https://doi.org/10.1007/b102410

Zsambok, C. E. (1997). Naturalistic decision making: Where are we now? In Z. C. E. & G. Klein (Eds.), *Naturalistic decision making* (pp. 3–16). Mahwah, NJ: Lawrence Erlbaum Associates, Inc.

9 Lenses of Diversity in Distributed Teams

T. C. Endsley, G. A. Macht, D. Engome Tchupo, C. Hammett, and J. R. S. Brownson

CONTENTS

INTRODUCTION

With the advent of the information technology age, distributed teams are quickly becoming the norm within many human operations across domains, due to the expanse of globalization, international industry, and more operations within compounded, multifaceted environments (e.g., military coalition teams, business and trade, manufacturing, spaceflight, international disaster recovery). In addition, in recent years, there has emerged a greater level of social awareness that identities are not fixed, and are indeed multifaceted (Valverde, Sovet, & Lubart, 2017); that there is a lack of representation of certain populations within research in the teams literature (Henrich, Heine, & Norenzayan, 2010; Endsley, Reep, McNeese, & Forster, 2015); and that this requires a re-examination of how and why some teams outperform others.

The establishment of common ground among team members is a fundamental component for the development of shared team cognition across all members of a team in situ. Team cognition is an emergent property that reflects "an emergent state [within a team] that refers to the manner in which knowledge important to team functioning is mentally organized, represented, and distributed within the team and allows team members to anticipate and execute actions (Kozlowski & Ilgen, 2006)" (DeChurch & Mesmer-Magnus, 2010, p. 33). In distributed teamwork environments, teams may be carrying out activities in which they are geographically, temporally, or virtually displaced, which can impact the ways that communication, collaboration, and coordination are carried out within a team, and which can greatly reduce a team's ability to develop common ground among team members (Hinds & Mortensen, 2005; Mancuso & McNeese, 2012; Powell, Piccoli, & Ives, 2004; Rosen, Furst, & Blackburn, 2007).

We trace team diversity here as an ensemble of individual differences, expressed collectively through shared spaces of belonging and shared purpose (i.e., joint intentionality among a group of individuals, such as a team). Differences for individual agents within a team can be the starting point for creativity, resilience, sustained performance, and project fulfillment. Greater levels of diversity within teams are often identified as a key mechanism for the avoidance of groupthink; engaging multiple perspectives of individual team members on task-specific issues, through positive conflict behaviors, has been shown to reduce team groupthink and improve team performance (Jehn, 1995; Kozlowski & Ilgen, 2006). With groupthink there is a tendency for a group to reach premature consensus on a particular issue, using shared cultural norms, values, and problem-solving heuristics, often to the detriment of robust problem solving, project fulfillment, and/or sustained team performance (Carnevale & Probst, 1998; Jehn, 1995; Kozlowski & Ilgen, 2006).

Diversity of individuals among teams can draw from unique intersectionalities expressed by individual human agents; who may also have diverse inhuman agents (e.g., AI and autonomous/robotic systems) as teammates (as future human machine teaming ambitions have emerged). Intersectionality describes the entangled superposition of factors tracing identity and power differentials of individuals co-created among society and shared communities, such as factors of race, gender, sexual orientation, disability, social background, cognitive background, and stories of place (Crenshaw, 1989). Tapping into the collective insight of team diversity has been shown to be a key framework enabling and maintaining creative, resilient, and robust performance (Kozlowski & Ilgen, 2006).

How collective insight emerges in team settings, and what can be ascertained from these concepts of diversity influencing cognition, sets the stage for new areas of focus within the team cognition literature, particularly as we seek to understand why some teams succeed and others fail. For example, within the teaming literature, overall higher average levels of emotional intelligence (EI) on teams have been found to be a large predictor of team performance (Chang, Sy, & Choi, 2012; Côte, 2007; Elfenbein, 2006; Macht, Nembhard, & Leicht, 2019).

Multiple aspects of diversity will be explored within this chapter to answer the question of how diversity impacts the development of distributed team cognition. Answering this question within the context of team performance will provide perspective to this dynamic, evolving viewpoints that have permeated across fields of team literature. Several topics within the diversity literature that influence the success of teams will be discussed in terms of how these concepts impact the ways in which teams carry out their work and develop shared team cognition and shared mental models. Within this chapter, culture, race and ethnicity, cognitive styles, and personalities will be explored in the context of team functions and performance. These purposeful areas were selected as emerging topics that advance and complement the current literature, which is primarily on using individual factors or attributes as structural influences on teams.

CULTURE AND MULTINATIONAL TEAMWORK

Culture, as it shapes and structurally impacts cognition through an individual's experience through cultural behaviors and practices, provides a rich avenue by which to examine the aspects of cognitive differentiation among populations. Culture is described as shared knowledge, values, and norms that are "transmitted from one generation to the next, which . . . includes the knowledge, belief, art, law, morals, customs, and any other capabilities and habits acquired by man as a member of society," (D'Andrade, 1981, p. 179; Donald, 2000; Triandis, 2001). We apply the term *culture* to trace multiple dynamic experiences of quotidian life through common, intergenerational vernaculars and practices within a social group. Culture, used here, describes a particular set of beliefs, norms, and behaviors of a social group, but may also be applied to the context of organizational, political, or geolocated community (Connaughton & Shuffler, 2007). In many fields, membership to a nation-state is often used to prescribe culture to a social group, although communities often transcend national borders and incorporate ethnic, racial, and religious associations. However, within the literature, cultural studies often treat participant identities solely on descriptions of national membership (Connaughton & Shuffler, 2007; Earley & Mosakowski, 2000). In many ways, this approach can be limited to understanding the rich and dynamical influence of collective cultural experience in shaping cognitive and behavioral patterns of individuals, but has served as a proximal construct for researchers exploring the topic of culture in teams. Our treatment of culture in individuals and teams, however, deserves greater consideration.

Cultural processes of behaviors, norms, expression of dissent, and communication capabilities may present significant contextual boundaries for teams to navigate. Team cognition emerges in context and the ways in which teams create shared knowledge structures, such as a team mental model or a shared understanding, may be constructed based on the perspectives and expectations of the individual team members (DeChurch & Mesmer-Magnus, 2010). The development of shared situation awareness (SA) may be influenced significantly by the cultural context in which the team operates (Endsley, 1995; Strauch, 2010). As discussed by Strauch (2010), "differences in cognitive and perceptual styles can affect team performance by leading to differences in the way operators perceive and comprehend system cues, differences that can affect situation awareness and subsequent decision making." If cultural norms dictate that subordinate team members do not overtly challenge authority or leadership, they may not share information, which may be pertinent to the task at hand (McHugh, Smith, & Sieck, 2008; Strauch, 2010). For example, Asiana Airlines Flight 214, which crashed on a runway at San Francisco International Airport in July of 2013, presents a failure of a team to effectively respond to changes in their descent, which some researchers have attributed to a possible impact of the pilots' Korean cultural background, where challenges to leadership authority are uncommon (Howard, 2013).

Multinational teams simultaneously provide an urgent area in need of study as well as a rich context for examining and understanding the role culture plays on cognition and team cognitive processes. Cultural differences in processes, values, etc. most often emerge clearly in juxtaposition to other cultures and their processes of doing things. Additionally, multinational team composition requires management of conflict and the negotiation of shared team perspectives by all of the team members. Due to the extensive use of mixed cultural teams to carry out tasks in mission-critical operations (e.g., disaster response, medicine, engineering, business, space flight operations), it is important to understand the processes of multinational teams.

Multinational teams will, by the nature of their composition, have many unique perspectives and as a result, team cohesion may be automatically limited as team members polarize towards the group of their social identity and, importantly, against members that do not share their identity (Shokef & Erez, 2008; Tajfel, 1982). When teams are composed of members with radically different decision-making processes, the potential for subgroup formation based on contentions of what that culture values in the decision-making process increases, and leads to a reduction in team cognition, particularly when conflict is not navigated effectively (Lau & Murnighan, 2005; Carton & Cummings, 2012). A growing body of literature has focused on the issues and benefits that emerge from the use of multinational teams across operational contexts, as they have been increasingly used.

The effect of personal culture on dynamic team process is evidenced firsthand from accounts of coalitions teams across multiple domains. Coalition teams are often used in the battlefield theatre or in crisis situations (Phillips, Ting, & Demurjian, 2002). A coalition is an alliance of international cooperation of "governmental, military, civilian, and international organizations" (Phillips et al., 2002, p. 87). Working in coalitions is a pivotal aspect of U.S. operations and strategy (Phillips et al., 2002), and from reports of coalition teams in theater (i.e., military or NATO teams), anecdotal evidence presents a compelling case for the need to incorporate and include culture as an impacting factor on cognition, and importantly, on team performance (Poteet et al., 2009). Even cultures that may typically be seen as culturally close or similar on some scales of culture (i.e., Hofstede's dimensions), such as the U.S. and the UK, present dramatic differences in the field for coalition teams seeking to work together (Phillips et al., 2002; Poteet et al., 2009). During operations these teams felt the effects and the "repercussions of substantive differences," indicating that a deeper look into the role of culture, definitions and overall scope/depth of impact of culture on cognition may need to be re-evaluated as a part of team interactions (Rasmussen, Sieck, & Smart, 2009).

It is often through interactions within multinational teams that unknown information about culturally defined team processes emerges, as cultural style and conflict and must be overcome to achieve team or mission goals. Poteet et al. (2009, p. 4) point out two important findings from their study on British and American military coalition teams: "(1) it is important to look at miscommunication in 'context', and (2) it is crucial to have a shared common understanding of the context." It is doubly important from an examination of these coalition teams to realize the role that a shared understanding of context plays in team performance and the ways in which culture emerges as a potential factor in influencing the development of common ground (Poteet et al., 2009).

In multinational teams, aspects of cultural differences seem to create issues for the development of shared team mental models (Carton & Cummings, 2012), particularly when members align on subgroup fronts, where they are unlikely to engage in appropriate coordination activities, or simply not know that they are missing anything at all. In multinational teams, the nature of team composition can have an enormous effect on group processes and outcomes, in which heterogeneity can affect trust, group commitment, communication, and cohesion (Klein et al., 2019; Polzer, Crisp, Jarvenpaa, & Kim, 2006).

Additionally, team situation awareness may be affected by attention differences among cultural populations, and what is viewed as a priority during team operation processes. If the focus of attention across cultural populations isn't shared cognitively (i.e., where an individual from a "collectivist culture" views a scenario "holistically" and another from an "individualistic culture" views only specific features), critical information could mistranslate or be missed (Kitayama & Uskul, 2011). It is possible, given this finding that there could be a potential outcome on team situation awareness and on overall team cognition, however, the direct outcomes of these cognitive differences will need to be explored empirically (Strauch, 2010).

RACIO-ETHNIC DIVERSITY IN TEAMS

The "Workforce 2000" report (Johnston & Packer, 1987) forecast that there would be significant increases in the number of women and minority groups entering the workforce by the year 2000. This report generated considerable interest in workforce diversity and its effect on teams and performance, leading to a significant body of research. While some research focused on both gender and racio-ethnic diversity (among other demographics), many have focused on just gender or racio-ethnic diversity. A recent census of the U.S. workforce shows that women make up 40% of the workforce while non-whites make up 22% of the workforce (Litaker & Bell, 2019).

While the literature generally accepts that diversity influences teams (Williams & O'Reilly, 1998), research has yielded inconsistent findings of the effects of racio-ethnic diversity on teams (Baugh & Graen, 1997; Harrison, Price, & Bell, 1998). Of the different studies, some results suggest that racio-ethnically homogeneous teams are more cohesive, as heterogeneous teams experience more conflict than homogeneous ones (Tsui, Egan, & O'Reilly, 1992; Milliken & Martins, 1996; Barsade & Gibson, 1998). These conflicts were attributed to the fact that visible differences can lead to reactions due to prejudices (Tsui et al., 1992; Milliken & Martins, 1996).

Rock, Grant, and Grey (2016) argue that while diversity does increase conflict, people tend to exaggerate the amount of conflict existing in diverse teams. They defend this view by pointing to a 2015 study (Lount, Sheldon, Rink, & Phillips, 2015) in which MBA students were told to imagine themselves managing several teams of interns, with one team requesting additional resources. The researchers gave the MBA students the same transcript of a conversation between the team members; the only difference was in the picture provided in the transcript. They were either given an image of four black men, four white men, or two black and two white men. The results of this study showed that the racially homogeneous teams were perceived as having the same level of interpersonal conflict while the heterogeneous teams were perceived as having higher levels of interpersonal conflict, and were subsequently less likely to be allocated the additional resources (Lount et al., 2015). The results indicate more bias and exaggeration in the perception of conflict than in reality.

More recently, as a result of the advances in communication technology and in an effort to mitigate the (perceived) conflicts created from visible differences, research has looked into the effect of communication media on the impact of race (and gender) in teams (Robert, Dennis, & Ahuja, 2018). One of the findings was that the use of text communication helps teams overcome issues associated with their racial diversity. The experiment found that for the teams that used text communication, racial diversity was positively associated with knowledge sharing and integration. This result supports others that claim racio-ethnic diversity may improve the quality of creative brainstorming (McLeod, Lobel, & Cox, 1998; Williams & O'Reilly, 1998). Overall, this provides ample opportunities for productive and effective communication in distributed teams via text, such as online work systems (e.g., Slack.com and Monday.com).

Despite the conflicts that may arise in racially heterogeneous teams, research suggests that one way of avoiding these conflicts and taking advantage of team diversity is to highlight the value of multiculturalism. Although the concept of multiculturalism (which deals with theories of differences) in postcolonial countries had previously been rejected by anti-racist groups in countries like Great Britain (Hall, 1995; Gunew, 1997), a 2009 study which had pairs of students (one white Canadian and one Aboriginal Canadian) team up for a conversation found that prefacing meetings with a message supporting multiculturalism (versus no message) was viewed more positively (by the students) than messages endorsing colorblindness, which led white students to become more negative toward their minority partners (Vorauer, Gagnon, & Sasaki, 2009).

Comprehensive literature reviews of several decades of research on this topic have concluded that while increased diversity at the micro-level (within groups) typically has an adverse effect on the team's ability to meet its members' needs and to function effectively over time (Williams & O'Reilly, 1998), there are no consistent main effects of demographic diversity on performance (Mohammed & Angell, 2004; Roberson, 2019). These conclusions suggest that any inconsistencies regarding the results of the impact of racio-ethnic diversity on team performance reflect the fact that there exist variables modifying these effects of racio-ethnic diversity on teams that are not being considered (Milliken & Martins, 1996; Pelled, Eisenhardt, & Xin, 1999).

One agreed-upon variable affecting the performance of diverse racio-ethnic teams is the context in which the team operates, coined in team literature as the environment (Milliken & Martins, 1996; Williams & O'Reilly, 1998; Riordan, 2000; Martins, Milliken, Wiesenfeld, & Salgado, 2003). Specifically, the racio-ethnic diversity of a group affects the group differently depending on if it functions within a similarly racio-ethnically diverse organization or not. Research by Martins et al. (2003) explains that in an environment with little racio-ethnic diversity, racio-ethnic diversity within a group/team has an adverse effect on team members' experiences as compared to a racio-ethnically diverse team in an equally diverse environment. Furthermore, for a racio-ethnically diverse team within an equally diverse environment, it is deep-level diversity, collectivism in this case, that harms group members' experiences. Collectivism was not found to have an effect on teams within a homogeneous context. Further results support that numeric minorities within organizations suffer from increased performance pressure, isolation from social and professional networks, and stereotypical role encapsulation (Kanter, 1977; Wharton, 1992). In addition, while members of heterogeneous teams (gender and race) may experience working relationships equal to those of homogeneous teams, they feel they must work harder to create and maintain them (Ely, 1994, 1995). Supervisors, meanwhile, were found to provide higher performance ratings to subordinates of the same race as them (Kraiger & Ford, 1985).

Despite all the work done in this field, more remains. One of the challenges with research on the effects of race and diversity in teams is that much of the research has taken place within disconnected research traditions. Human factors, as it is inherently interdisciplinary, could provide tremendous contributions toward aggregating these different fields in order to better understand diversity in teams. A review of recent literature (Roberson, 2019) identified several aspects of diversity in which research is still lacking. One presented avenue is utilizing temporal or dynamic approaches to account for demographic shifts over time. In addition, more research is needed on investigating the effects of race on interactional behaviors and patterns within work units, understanding the "P" in race Input-Process-Output (I-P-O) models, and studying race and diversity in nonbusiness contexts. The literature is also lacking a focus on the individuals making up the teams. Little is known about the effects of racio-ethnic team diversity on the individual members of the team, as most studies focus on team outcomes. Furthermore, many studies look at performance as an outcome where it would be beneficial to consider other group and individual level outcomes, such as conflict.

COGNITIVE STYLE

Individual diversity in how people think, process, believe, and comprehend information can fall beneath initial observation (Harrison et al., 1998; Mannix & Neale, 2005; Aggarwal, Woolley, Chabris, & Malone, 2019). Empirically and biologically, for example, information formats are processed differently by different people (Richardson, 1977; Cabeza & Nyberg, 2000). The success of comprehending information often depends on the intersection of its presented format and an individual's mental processing, or cognitive style. Cognitive style, whose origins emerged in the 1940s, defines how quickly and accurately an individual understands conceptual information, how they frame problems, and further, how they behave during decision making (Mello & Delise, 2015).

Understanding a team's composition of cognitive styles can provide insight into the success of distributed teams. As pointed out by Golian (1998), our thinking style affects how we relate to others, reason, solve problems, and communicate. This provides foresight into forming well-integrated teams. Connecting teams with a diversity of cognitive strengths introduces differences in skills and perspectives, which according to Aggarwal and Woolley (2019) are the raw ingredients to inspire creativity and reveal a high potential for effective team performance (De Dreu & West, 2001; Sung & Choi, 2012; Aggarwal & Woolley, 2019). Cognitive diversity allows for a variety of information processing approaches, leading to differing decisions and behaviors. Incorporating a variety of these qualities into a team can create pathways for creative and structured ideas that may not be explored at an individual cognitive level.

Cognitive style has been studied along multiple constructs for individuals, but in teams it has primarily been conceptualized in terms of verbal, spatial-, and object-visualization construct (Schilpzand, 2010; Aggarwal & Woolley, 2013, 2018; Aggarwal et al., 2019). Based on this approach, object-visualization uses holistic processing, whereas spatial-visualization uses spatial relationships, and verbalization uses strategies of linguistic analysis (Bartlett, 1932; Paivio, 1971; Richardson, 1977; Kozhevnikov, Kosslyn, & Shephard, 2005; Kozhevnikov, Evans, & Kosslyn, 2014; Aggarwal et al., 2019). By creating an information format that resonates with the trio of possible individual cognitive styles, people's intake of information and its adjoined retrieval can be improved, and subsequent team-based decision making better modeled.

Two opposing viewpoints are explained by the information-processing perspective and the socially shared cognition perspective (Aggarwal et al., 2019). Socially shared cognition argues that diverse teams have a hard time understanding each other, while information processing insists diversity enables new viewpoints of thought (Amabile, 1983; Cox & Blake, 1991; De Dreu & West, 2001; Homan, Buengeler, Eckhoff, Ginkel, & Voelpel, 2015; Van Knippenberg & Schippers, 2007; Aggarwal & Woolley, 2019). More research is needed to explore how socially shared cognition and information processing can work together to form balanced, interactive teams in distributed settings and other areas of collaboration, however current research provides evidence to support both propositions.

The most recent research on the diversity of cognitive style in teams shows a negative effect on strategic consensus, an indirect positive effect on team creativity and an area of peak performance with collective intelligence (i.e., the aggregated intelligence of individuals interacting in a team setting) which then declines when the spread of cognitive diversity becomes too large for mutual understanding of intelligence (Aggarwal et al., 2019; Aggarwal & Woolley, 2019). This supports both cognitive perspectives described by Aggarwal et al. (2019) and gives scope to the thought that there is a specific placement of balance to cognitive diversity in team composition to reach high team performance. Uncovering the factors that contribute to the understanding of opposing cognitive styles in teams will ultimately aid in the understanding of cognitive variance and connect the diversity of cognitive styles to high team performance (Mello & Delise, 2015).

PERSONALITY

Personality traits and an individual's cognitive style are inextricably linked as they are both integral paradigms through which individuals react to the world, forming a human processing lens (McGhee, Shields, & Birnberg, 1978; Pratt, 1980). While there have been likenesses identified, individuals with the same or similar personalities may intake and process information differently and arrive at different conclusions from their counterparts (McGhee et al., 1978). Research relevant to how personality affects team performance can be discussed in terms of the effects of cognitive style and emotional intelligence (EI).

Cognitive Style and Personality

A majority of the work exploring the overlap between personality and cognitive style exists at the individual level (Rothstein & Goffin, 2006; John, Naumann, & Soto, 2008; Duff, Boyle, Dunleavy, & Ferguson, 2004; Gellatly, 1996; Joseph & Newman, 2010). However, there is no clear, distinguishable overlap between the personality or cognitive style constructs with respect to teams.

Sprehn, Macht, Okudan, and Nembhard (2013) assert that if personality and cognitive style relate to one another, it would be between the Five-Factor Model's (FFM) conscientiousness and three-dimension of cognitive style (i.e., verbal, spatial-, and object-visualizer). Sprehn et al. (2013) explored these complex relationships in the presence of a team setting and indicated a direct link between a team's mean measure of conscientiousness and team performance. While there was indeed a mediating relationship between those two variables based on the cognitive diversity within the object-visualizer component of the team, it was a negative relationship. This aids to the claim of the socially shared cognition perspective (i.e., a negative effect of cognitive style diversity), and reiterates the need to explore how diverse teams can effectively share information that is perceived differently to create collectively understood team mental models and success. Further expansion of this concept needs to be explored to fully understand the overall effects of how a teams' personality can influence its collective cognitive style (i.e., cognitive styles of individuals interacting in a team setting) in order to understand the implications of team diversity for high team performance in both distributed and collocated teams.

Emotional Intelligence, Personality, and Team Performance

Within the past several decades, research has attempted to understand the relationship between personality and team performance (Morgeson, Reider, & Campion, 2005; Kozlowski & Bell, 2003; Rothstein & Goffin, 2006) as well as emotional intelligence (EI) and team performance (Chang et al., 2012; Leicht, Macht, Riley, & Messner, 2013; Macht et al., 2019), separately. Even with advanced statistical tools, it is difficult to fully predict team performance based on their separate metrics, let alone together (Hough, 2001).

A self-reported trait, EI has components of learning (Thompson, 2009) with measurements that significantly overlap with the Five-Factor Model and general intelligence (Joseph & Newman, 2010; O'Boyle, Humphrey, Pollack, Hawver, & Story, 2011), which could provide insight into people's collaborative potential as it affects team performance. EI explains how individuals understand and potentially control their emotional response to daily experiences, something that is critical to social interactions (Bar-On, 2006; Bar-On, 1997; Mayer & Salovey, 1997; Goleman, 1995). Researchers claim that higher average levels of EI in teams allow them to achieve better collective team decisions and still be more adaptable to team behaviors while elevating coordination and effective communication (Chang et al., 2012; Côte, 2007; Elfenbein, 2006; Macht et al., 2019). The team collective EI has also been related to team leadership (Chang et al., 2012), leadership emergence (Côte, Lopes, Salovey, & Miners, 2010), and team task orientation and team maintenance (Frye, Bennett, & Caldwell, 2006), but only in the beginning stages of a team project (Jordan, Ashkanasy, Härtel, & Hooper, 2002). Relating team performance to traditional team measures, individual-level, team-level, and environmental-level factors is required not only to understand how EI relates to team performance but also to help understand a team's inherently complex relationships (Driskell, Hogan, & Salas, 1987).

Zeidner (1995) presented seven potential types of relationships that could exist between personality and intelligence constructs; however, researchers have since agreed that a few of these proposed relationships could be too simplistic. Moderation and mediation relationship frameworks have been researched for personality and performance (Rothstein & Goffin, 2006; Peeters, van Tuijl, Rutte, & Reymen, 2006a; Peeters, Rutte, van Tuijl, & Reymen, 2006b; Stewart, 2006; Graziano, Hair, & Finch, 1997; Bond & Ng, 2004; Ployhart & Ehrhart, 2003; Heller & Watson, 2005; Arvey, Rotundo, Johnson, Zhang, & McGue, 2006), and a few works address EI and performance in this way (Chang et al., 2012; Rode et al., 2007; Farh, Seo, & Tesluk, 2012). Limited work explores these statistical relationships at the team level for personality and team performance (Macht & Nembhard, 2015; Macht, Nembhard, Kim, & Rothrock, 2014), however, and EI and team performance (Macht, 2014).

Macht (2014) demonstrated that personality and EI did indeed impact each other in different ways based on the task type, through either moderating effects or mediating effects. Among team studies, there are two broad categories of research that empirically examine either moderating effects (i.e., pathways in which the context and environment overshadow [e.g., personality] the overall effect that independent variables [e.g., EI] have on the dependent variable [e.g., team performance]) or mediating effects (i.e., pathways through [e.g., personality] which the independent variable [e.g., EI] being studied affects the dependent variable [e.g., team performance]) (Baron & Kenny, 1986; Horwitz, 2005; Rothstein & Goffin, 2006). Personality and EI moderated one another during intellectual/analytical tasks (i.e., "generation, exploration, or verification of knowledge" [Driskell et al., 1987, p. 104]) and logical/precision tasks (i.e., "performance of explicit, routine tasks or task requiring attention to detail" [Driskell et al., 1987, p. 104]), while a mediating relationship was indicated only during the logical/precision task. These results show that there are many nuances to the implementation and interpretation of these complex relationships, significantly hindering understanding for team performance.

Despite the complexity of Macht's (2014) models, the results reflect an unforeseen relationship between the interactions of personality and EI in the prediction of team performance which are dependent not only on the level of interaction but also on the task type. In order to explore the internal complexities between the EI and personality metrics, and their relationships to team performance, advanced statistical methods must be incorporated. Knowing their potential relationships and how this affects knowledge sharing, collaboration, and decision making in teams, however, could provide useful predictors of team performance (Furnham, 2008), especially within highly complex tasks and distributed teams. With future research, it could be possible to create a combination of individual EI and personality traits that would be indicative of high-performance collaboration in teams.

CONCLUSION

This chapter explores how nuances in individual identities should not be overlooked and that individual differences among team members should not be discounted when talking about how teams perform in distributed contexts. The influence of diverse perspectives on team performance outcomes implicates a different nuance to team behaviors and interactions that deserves attention and further examination, as the very nature of teamwork is shifting to incorporate composition of highly diverse cultural, racio-ethnic, and cognitive styles and is extending to include multiple kinds of agents with different personalities and collective and emotional intelligences. Knowing how various levels of identity demographics (e.g., culture, racio-ethnicity) and cognitive metrics (e.g., styles, personality, emotional intelligence) do or do not relate to team performance is vital to fully comprehending how to form teams in a more universal, generalizable manner. Approaches from a variety of fields provide unique opportunities for extension into teams research. In more recent years, for example, research focus has shifted from looking at one aspect of diversity (e.g., race, gender) to look at the intersectionality (Crenshaw, 1985) of diverse characteristics within individuals. Intersectionality has been largely unexplored as it relates to team dynamics. Capturing multifaceted and/or multilayered nature of intra-individual identity (Roberson, 2019) as it relates to team performance outcomes is an important next step; intersectionality may provide a new lens through which to understand team processes. Extending diversity research in distributed teams is a critical next step as many areas of diverse team composition, as their interactions as constructs within teams, are not well explored.

Ultimately, teaming requires a collection of individuals (e.g., humans, intelligent agents, robots) to coalesce into a teams' greater functional capabilities. As Salas, Cooke, and Rosen (2008) suggest, "Teams are used when errors lead to severe consequences; when the task complexity exceeds the capacity of an individual; when the task environment is ill-defined, ambiguous, and stressful; when multiple and quick decisions are needed; and when the lives of others depend on the collective insight of individual members" (p. 540). Understanding how culturally, socially, and cognitively diverse teams can operate successfully (i.e., achieving goals and interacting effectively to do so) is quintessential given the employment of diverse teams in a variety of operational contexts.

REFERENCES

Aggarwal, I., & Woolley, A. W. (2013). Do you see what I see? The effect of members' cognitive styles on team processes and performance. *Organizational Behavior and Human Decision Processes*, 122, 92–99.

Aggarwal, I., & Woolley, A. W. (2018). Team creativity, cognition, and cognitive style diversity. *Management Science*, *65*(4), 1586–1599. doi: 10.1287/mnsc.2017.3001

Aggarwal, I., & Woolley, A. W. (2019). Team creativity, cognition, and cognitive style diversity. *Management Science*, *65*(4), 1586–1599. https://doi.org/10.1287/mnsc.2017.3001.

Aggarwal, I., Woolley, A. W., Chabris, C. F., & Malone, T. W. (2019). The impact of cognitive style diversity on implicit learning in teams. *Frontiers In Psychology*, 10. https://doi.org/10.3389/fpsyg.2019.00112.

Amabile, T. M. (1983). The social psychology of creativity: A componential conceptualization. *Journal of Personality and Social Psychology*, *45*(2), 357–376.

Arvey, R. D., Rotundo, M., Johnson, W., Zhang, Z., & McGue, M. (2006). The determinants of leadership role occupancy: Genetic and personality factors. *The Leadership Quarterly*, *17*(1), 1–20.

Bar-On, R. (1997). *The Emotional Quotient Inventory (EQ-i): A test of emotional intelligence*. Toronto, Canada: Multi-Health Systems.

Bar-On, R. (2006). The Bar-On model of Emotional-Social Intelligence (ESI). *Psicothema*, *18*, 13–25.

Baron, R. M., & Kenny, D. A. (1986). The moderator–mediator variable distinction in social psychological research: Conceptual, strategic, and statistical considerations. *Journal of Personality and Social Psychology*, *51*(6), 1173–1182. doi: 10.1037/0022-3514.51.6.1173

Barsade, S. G., & Gibson, D. E. (1998). Group emotion: A view from top and bottom. In D. H. Gruenfeld (Ed.), *Composition* (pp. 81–102). Greenwich, CT: JAI Press.

Bartlett, F. C. (1932). *Remembering: A study in experimental and social psychology*. Cambridge: Cambridge University Press.

Baugh, G. S., & Graen, G. B. (1997). Effects of team gender and racial composition on perceptions of team performance in cross-functional teams. *Group & Organization Management*, *22*(3), 366–383.

Bond, M. H., & Ng, I. W.-C. (2004). The depth of a group's personality resources: Impacts on group process and group performance. *Asian Journal of Social Psychology*, *7*, 285–300.

Cabeza, R., & Nyberg, L. (2000). Imaging cognition II: An empirical review of 275 PET and fMRI studies. *Journal of Cognitive Neuroscience*, *12*(1), 1–47. Retrieved from www.ncbi.nlm.nih.gov/pubmed/10769304.

Carnevale, P. J., & Probst, T. M. (1998). Social values and social conflict in creative problem solving and categorization. *Journal of Personality and Social Psychology*, *74*(5), 1300.

Carton, A. M., & Cummings, J. N. (2012). A theory of subgroups in work teams. *Academy of Management Review*, *37*(3), 441–470.

Chang, J. W., Sy, T., & Choi, J. N. (2012). Team emotional intelligence and performance: Interactive dynamics between leaders and members. *Small Group Research*, *43*(1), 75–104.

Connaughton, S. L., & Shuffler, M. (2007). Multinational and multicultural distributed teams: A review and future agenda. *Small Group Research*, *38*(3), 387–412. https://doi.org/10.1177/1046496407301970.

Côte, S. (2007). Group emotional intelligence and group performance. In *Affect and groups* (pp. 309–336). Bingley: Emerald Group Publishing Limited.

Côte, S., Lopes, P. N., Salovey, P., & Miners, C. T. H. (2010). Emotional intelligence and leadership emergence in small groups. *The Leadership Quarterly*, *21*(3), 496–508.

Cox, T. H., & Blake, S. (1991). Managing cultural diversity: Implications for organizational competitiveness. *Academy of Management Perspectives*, *5*(3), 45–56. doi: 10.5465/ame.1991.4274465

Crenshaw, K. (1989). Demarginalizing the intersection of race and sex: A Black feminist critique of anti-discrimination doctrine, feminist theory and anti-racist politics. *University of Chicago Legal Forum*, *1*, 139–167.

Crenshaw, M. (1985). An organizational approach to the analysis of political terrorism. *Orbis*, *29*, 465–489.

D'Andrade, R. G. (1981). The cultural part of cognition. *Cognitive Science*, *5*, 179–195. https://doi.org/10.1207/s15516709cog0503_1.

De Dreu, C. K., & West, M. A. (2001). Minority dissent and team innovation: The importance of participation in decision making. *Journal of Applied Psychology*, *86*(6), 1191–1201.

DeChurch, L. A., & Mesmer-Magnus, J. R. (2010). The cognitive underpinnings of effective teamwork: A meta-analysis. *Journal of Applied Psychology, 95*(1), 32–53. https://doi.org/10.1037/a0017328

Donald, M. (2000). The central role of culture in cognitive evolution: A reflection on the myth of the isolated mind. In L. Nucci, G. Saxe, & E. Turiel (Eds.), *Culture, thought and development*. Mahwah, NJ: Lawrence Erlbaum Associates.

Driskell, J. E., Hogan, R., & Salas, E. (1987). Personality and group performance. In C. Hendrick (Ed.), *Review of personality and social psychology* (pp. 91–112). Newbury Park, CA: Sage.

Duff, A., Boyle, E., Dunleavy, K., & Ferguson, J. (2004). The relationship between personality, approach to learning and academic performance. *Personality and Individual Differences, 36*, 1907–1920.

Earley, P. C., & Mosakowski, E. (2000). Creating hybrid team cultures: An empirical test of transnational team functioning. *Academy of Management Journal, 43*(1), 26–49.

Elfenbein, H. A. (2006). Team emotional intelligence: What it can mean and how it can impact performance. In V. Druskat, F. Salas, & G. Mount (Eds.), *The link between emotional intelligence and effective performance* (pp. 165–184). Mahwah, NJ: Lawrence Erlbaum.

Ely, R. J. (1994). The effects of organizational demographics and social identity on relationships among professional women. *Administrative Science Quarterly, 39*(2), 203. doi: 10.2307/2393234

Ely, R. J. (1995). The power in demography: Women's social constructions of gender identity at work. *Academy of Management Journal, 38*(3), 589–634. doi: 10.2307/256740

Endsley, M. R. (1995). Measurement of situation awareness in dynamic systems. *Human Factors, 37*(1), 65–84.

Endsley, T. C., Reep, J. A., McNeese, M. D., & Forster, P. K. (2015, September). Conducting cross national research: lessons learned for the human factors practitioner. In *Proceedings of the Human Factors and Ergonomics Society annual meeting* (Vol. 59, No. 1, pp. 1147–1151). Los Angeles, CA: SAGE Publications.

Farh, C. I. C. C., Seo, M.-G., & Tesluk, P. E. (2012). Emotional intelligence, teamwork effectiveness, and job performance: The moderating role of job context. *Journal of Applied Psychology, 97*(4), 890–900.

Frye, C. M., Bennett, R., & Caldwell, S. (2006). Team emotional intelligence and team interpersonal process effectiveness. *Mid-American Journal of Business, 21*(1), 49–56.

Furnham, A. (2008). *Personality and intelligence at work: Exploring and explaining individual differences at work*. New York: Routledge.

Gellatly, I. R. (1996). Conscientiousness and task performance: Test of cognitive process model. *Journal of Applied Psychology, 81*(5), 474–482. https://doi.org/10.1037/0021-9010.81.5.474

Goleman, D. (1995). *Emotional intelligence: Why it can matter more than IQ*. New York: Bantam Books.

Golian, L. M. (1998). *Thinking style differences among academic librarians*. Boca Raton, FL: Florida Atlantic University.

Graziano, W. G., Hair, E. C., & Finch, J. F. (1997). Competitiveness mediates the link between personality and group performance. *Journal of Personality and Social Psychology, 73*(6), 1394–1408.

Gunew, S. (19970. Postcolonialism and multiculturalism: between race and ethnicity. *The Yearbook of English Studies, 27*, 22. https://doi.org/10.2307/3509130.

Hall, S. (1995). New ethnicities. In B. Ashcroft, G. Griffiths, & H. Tiffin (Eds.), *The postcolonial studies reader*. Abingdon: Routledge.

Harrison, D. A., Price, K. H., & Bell, M. P. (1998). Beyond relational demography: Time and the effects of surface- and deep-level diversity on work group cohesion. *Academy of Management Journal*, *45*(5), 1029–1045.

Heller, D., & Watson, D. (2005). The dynamic spillover of satisfaction between work and marriage: The role of time and mood. *Journal of Applied Psychology*, *90*(6), 1273–1279.

Henrich, J., Heine, S. J., & Norenzayan, A. (2010). The weirdest people in the world? *Brain and Behavioral Sciences*, *33*, 61–135. https://doi.org/10.1017/S0140525X0999152X

Hinds, P. J., & Mortensen, M. (2005). Understanding conflict in geographically distributed teams: The moderating effects of shared identity, shared context, and spontaneous communication. *Organization Science*, *16*(3), 290–307. https://doi.org/10.1287/orsc.1050.0122

Homan, A. C., Buengeler, C., Eckhoff, R. A., Ginkel, W. P. V., & Voelpel, S. C. (2015). The interplay of diversity training and diversity beliefs on team creativity in nationality diverse teams. *Journal of Applied Psychology*, *100*(5), 1456–1467. doi: 10.1037/apl0000013

Horwitz, S. K. (2005). The compositional impact of team diversity on performance: Theoretical considerations. *Human Resource Development Review*, *4*(2), 219–245. doi: 10.1177/1534484305275847

Hough, L. M. (2001). I/owes its advances to personality. In B. W. Roberts & R. Hogan (Eds.), *Personality psychology in the workplace*. Washington, DC: American Psychological Association.

Howard, B. C. (2013). Could Malcolm Gladwell's theory of cockpit culture apply to Asiana crash? *National Geographic*. Retrieved from http://news.nationalgeographic.com/news/2013/07/130709-asiana-flight-214-crash-korean-airlines-culture-outliers/.

Jehn, K. A. (1995). A multimethod examination of the benefits and detriments of intragroup conflict. *Administrative Science Quarterly*, 256–282.

John, O. P., Naumann, L. P., & Soto, C. J. (2008). Paradigm shift to the integrative big five trait taxonomy: History, measurement, and conceptual issues. In O. P. John, O. P., L. P. Naumann, & C. J. Soto (Eds.), *Handbook of personality: Theory and research* (Vol. 3). New York: The Guilford Press.

Johnston, W. B., & Packer, A. E. (1987). *Workforce 2000: Work and workers for the 21st century*. Indianapolis, IN: Hudson Institute.

Jordan, P. J., Ashkanasy, N. M., Härtel, C. E. J., & Hooper, G. S. (2002). Workgroup emotional intelligence scale development and relationship to team process effectiveness and goal focus. *Human Resource Management Review*, *12*, 195–214.

Joseph, D. L., & Newman, D. A. (2010). Emotional intelligence: An integrative meta-analysis and cascading model. *Journal of Applied Psychology*, *95*(1), 54–78.

Kanter, R. M. (1977). *Men and women of the corporation*. New York: Basic Books.

Kitayama, S., & Uskul, A. K. (2011). Culture, mind, and the brain: Current evidence and future directions. *Annual Review of Psychology*, *62*, 419–449. https://doi.org/10.1146/annurev-psych-120709-145357.

Klein, H. A., Lin, M. H., Miller, N. L., Militello, L. G., Lyons, J. B., & Finkeldey, J. G. (2019). Trust across culture and context. *Journal of Cognitive Engineering and Decision Making*, *13*(1), 10–29.

Kozhevnikov, M., Evans, C., & Kosslyn, S. M. (2014). Cognitive style as environmentally sensitive individual differences in cognition a modern synthesis and applications in education, business, and management. *Psychological Science in the Public Interest*, *15*, 3–33. https://doi.org/10.1177/1529100614525555

Kozhevnikov, M., Kosslyn, S., & Shephard, J. (2005). Spatial versus object visualizers: A new characterization of visual cognitive style. *Memory and Cognition*, *33*(4), 710–726.

Kozlowski, S. W. J., & Bell, B. S. (2003). Work groups and teams in organizations. In W. C. Borman, D. R. Ilgen, & R. J. Klimoski (Eds.), *Handbook of psychology: Industrial and organizational psychology* (Vol. 12, pp. 333–375). New York: Wiley-Blackwell.

Kozlowski, S. W., & Ilgen, D. R. (2006). Enhancing the effectiveness of work groups and teams. *Psychological Science in the Public Interest*, *7*(3), 77–124.

Kraiger, K., & Ford, J. K. (1985). A meta-analysis of rater race effects in performance ratings. *Journal of Applied Psychology*, *70*, 56–65.

Lau, D. C., & Murnighan, J. K. (2005). Interactions within groups and subgroups: The effects of demographic faultlines. *Academy of Management Review*, *48*(4), 645–659.

Leicht, R. M., Macht, G. A., Riley, D. R., & Messner, J. I. (2013). Emotional intelligence provides indicators for team performance in an engineering course. *Engineering Project Organization Journal*, *3*(1), 2–12.

Litaker, G., & Bell, R. (2019). By the numbers: Diversity in the workplace. *Workforce.com*. Retrieved from www.workforce.com/2019/06/04/by-the-numbers-diversity-in-the-workplace/.

Lount, R. B., Sheldon, O. J., Rink, F., & Phillips, K. W. (2015). Biased perceptions of racially diverse teams and their consequences for resource support. *Organization Science*, *26*(5), 1351–1364. https://doi.org/10.1287/orsc.2015.0994.

Macht, G. A. (2014). *Modeling psychometrics for team performance: Personality & emotional intelligence*. Doctoral dissertation, The Pennsylvania State University, University Park, PA.

Macht, G. A., & Nembhard, D. A. (2015). Measures and models of personality and their effects on communication and team performance. *International Journal of Industrial Ergonomics*, *49*, 78–89. https://doi.org/10.1016/j.ergon.2015.05.006

Macht, G. A., Nembhard, D. A., Kim, J. H., & Rothrock, L. (2014). Structural models of extraversion, communication, and team performance. *International Journal of Industrial Ergonomics*, *44*(1), 82–91. https://doi.org/10.1016/j.ergon.2013.10.007 [pdf]

Macht, G. A., Nembhard, D. A., & Leicht, R. M. (2019). Operationalizing emotional intelligence for team performance. *International Journal of Industrial Ergonomics*, *71*, 57–63. https://doi.org/10.1016/j.ergon.2019.02.007

Mancuso, V. F., & McNeese, M. D. (2012). Effects of integrated and differentiated team knowledge structures on distributed team cognition. *Proceedings of the Human Factors and Ergonomics Society Annual Meeting*, *56*(1), 388–392.

Mannix, E., & Neale, M. A. (2005). What differences make a difference? *Psychological Science in the Public Interest*, *6*(2), 31–55. doi: 10.1111/j.1529-1006.2005.00022.x

Martins, L. L., Milliken, F. J., Wiesenfeld, B. M., & Salgado, S. R. (2003). Racioethnic diversity and group members' experiences: The role of the racioethnic diversity of the organizational context. *Group & Organization Management*, *28*(1), 75–106. https://doi.org/10.1177/1059601102250020

Mayer, J. D., & Salovey, P. (1997). What is emotional intelligence? In P. Salovey (Ed.), *Emotional development and emotional intelligence*. New York: Basic Books.

McGhee, W., Shields, M., & Birnberg, J. (1978). The effects of personality on a subject's information processing. *The Accounting Review*, 681–697.

McHugh, A. P., Smith, J. L., & Sieck, W. R. (2008). Cultural variations in mental models of collaborative decision making. In J. M. C. Schraagen, L. Militello, T. Ormerod, & R. Lipshitz (Eds.), *Macrocognition and naturalistic decision making*. Aldershot, UK: Ashgate Publishing Limited.

McLeod, P. L., Lobel, S. A., & Cox, T. H. (1998). Ethnic diversity and creativity in small groups. *Small Group Research*, *27*, 248–264.

Mello, A. L., & Delise, L. A. (2015). Cognitive diversity to team outcomes. *Small Group Research*, *46*(2), 204–226. https://doi.org/10.1177/1046496415570916.

Milliken, F. J., & Martins, L. L. (1996). Searching for common threads: Understanding the multiple effects of diversity in organizational groups. *Academy of Management Review*, *21*(2), 402–433.

Mohammed, S., & Angell, L. C. (2004). Surface-and deep-level diversity in workgroups: Examining the moderating effects of team orientation and team process on relationship conflict. *Journal of Organizational Behavior*, *25*(8), 1015–1039.

Morgeson, F. P., Reider, M. H., & Campion, M. A. (2005). Selecting individuals in team settings: The importance of social skills, personality characteristics, and teamwork knowledge. *Personnel Psychology*, *58*(3), 583–611.

O'Boyle, E. H., Humphrey, R. H., Pollack, J. M., Hawver, T. H., & Story, P. A. (2011). The relation between emotional intelligence and job performance: A meta-analysis. *Journal of Organizational Behavior*, *32*(5), 788–818.

Paivio, A. (1971). *Imagery and verbal processes*. New York: Holt, Rinehart & Winston.

Peeters, M. A. G., Rutte, C. G., van Tuijl, H. F. J. M, & Reymen, I. M. M. J. (2006b). The big five personality traits and individual satisfaction with the team. *Small Group Research*, *37*(2), 187–211.

Peeters, M. A. G., van Tuijl, H. F. J. M., Rutte, C. G., & Reymen, I. M. M. J. (2006a). Personality and team performance: A meta-analysis. *European Journal of Personality*, *20*(5), 377–396.

Pelled, L. H., Eisenhardt, K. M., & Xin, K. R. (1999). Exploring the black box: An analysis of work group diversity, conflict, and performance. *Administrative Science Quarterly*, *44*, 1–28.

Phillips, J. C., Ting, T. C., & Demurjian, S. (2002). *Information sharing and security in dynamic coalitions*. Paper presented at The Seventh ACM Symposium on Access Control Models and Technologies, Monterey, CA.

Ployhart, R. E., & Ehrhart, M. G. (2003). Be careful what you ask for: Effects of response instructions on the construct validity and reliability of situational judgment tests. *International Journal of Selection and Assessment,11*, 1–16.

Polzer, J., Crisp, C. B., Jarvenpaa, S. L., & Kim, J. W. (2006). Extending the faultline model to geographically dispersed teams how colocated subgroups can impair group functioning. *Academy of Management Journal*, *49*(4), 679–692.

Poteet, S., Giammanco, C., Patel, J., Kao, A., Xue, P., & Whitely, I. (2009). *Miscommunications and context awareness*. Paper presented at the 3rd Annual Conference of the International Technology Alliance (ACITA), Maryland, USA.

Powell, A., Piccoli, G., & Ives, B. (2004). Virtual teams: A review of current literature and directions for future research. *The DATA BASE for Advances in Information Systems*, *35*(1), 6–36.

Pratt, J. (1980). The effects of personality on subject's information processing: A comment. *The Accounting Review*, *55*(3), 501–506.

Rasmussen, L. J., Sieck, W. R., & Smart, P. (2009). What is a good plan? Cultural variations in expert planners' concepts of plan quality. *Journal of Cognitive Engineering and Decision Making*, *3*(3), 228–249. https://doi.org/10.1518/155534309X474479.

Richardson, A. (1977). Verbalizer-visualizer: A cognitive style dimension. *Journal of Mental Imagery*, *1*, 109–126.

Riordan, C. M. (2000). Relational demography within groups: Past developments, contradictions, and new directions. In G. R. Ferris (Ed.), *Research in personnel and human resources management* (Vol. 19, pp. 131–173). Greenwich, CT: JAI Press.

Roberson, Q. M. (2019). Diversity in the workplace: A review, synthesis, and future research agenda. *Annual Review of Organizational Psychology and Organizational Behavior, 6*, 69–88.

Robert, L. P. Jr., Dennis, A. R., & Ahuja, M. K. (2018). Differences are different: Examining the effects of communication media on the impacts of racial and gender diversity in decision-making teams. *Information Systems Research, 29*(3), 1–29.

Rock, D., Grant, H., & Grey, J. (2016, September). Diverse teams feel less comfortable—And that's why they perform better. *Harvard Business Review*. Retrieved from https://hbr.org/2016/09/diverse-teams-feel-less-comfortable-and-thats-why-they-perform-better

Rode, J. C., Mooney, C. H., Arthaud-Day, M. L., Near, J. P., Baldwin, T. T., Rubin, R. S., & Bommer, W. H. (2007). Emotional intelligence and individual performance: Evidence of direct and moderated effects. *Journal of Organizational Behavior, 28*(4), 399–421.

Rosen, B., Furst, S., & Blackburn, R. (2007). Overcoming barriers to knowledge sharing in virtual teams. *Organizational Dynamics, 36*(3), 259–273. https://doi.org/10.1016/j.orgdyn.2007.04.007

Rothstein, M. G., & Goffin, R. D. (2006). The use of personality measures in personnel selection: What does current research support? *Human Resource Management Review, 16*(2), 155–180.

Salas, E., Cooke, N. J., & Rosen, M. A. (2008). On teams, teamwork, and team performance: Discoveries and developments. *Human Factors, 50*(3), 540–547. https://doi.org/10.1518/001872008X288457

Schilpzand, M. C. (2010). *Cognitive diversity and team performance: The roles of team mental models and information processing mechanisms*. College of Management, Georgia Institute of Technology.

Shokef, E., & Erez, M. (2008). Cultural intelligence and global identity in multicultural teams. In S. Ang & L. Van Dyne (Eds.), *Handbook of cultural intelligence: Theory, measurement, and applications*. Armonk, NY: M.E. Sharpe, Inc.

Sprehn, K. A., Macht, G. A., Okudan, G. E., & Nembhard, D. A. (2013). Personality, cognitive style, and team performance. In *International Conference on Engineering Design (ICED) 2013*, August 19–22. Seoul, Korea: The Design Society.

Stewart, G. L. (2006). A meta-analytic review of relationships between team design features and team performance. *Journal of Management, 32*(1), 29–55.

Strauch, B. (2010). Can cultural differences lead to accidents? Team cultural differences and sociotechnical system operations. *Human Factors, 52*(2), 246–263. https://doi.org/10.1177/0018720810362238

Sung, S. Y., & Choi, J. N. (2012). Effects of team knowledge management on the creativity and financial performance of organizational teams. *Organizational Behavior and Human Decision Processes, 118*(1), 4–13. doi: 10.1016/j.obhdp.2012.01.001

Tajfel, H. (1982). Social psychology of intergroup relations. *Annual Review of Psychology, 33*, 1–39.

Thompson, H. L. (2009). Using the EQ-I and MSCEIT in tandem. In M. Hughes, H. L. Thompson, & J. B. Terrell (Eds.), *Handbook for developing emotional and social intelligence*. Hoboken, NJ: Wiley.

Triandis, H. C. (2001). Individualism-collectivism and personality. *Journal of Personality, 69*(6), 907–924. https://doi.org/10.1111/1467–6494.696169

Tsui, A. S., Egan, T. D., & O'Reilly, C. A. (1992). Being different: Relational demography and organizational attachment. *Administrative Science Quarterly, 37*, 549–579.

Valverde, J., Sovet, L., & Lubart, T. (2017). Self-construction and creative "life design". In M. Karwowski & J. C. Kaufman (Eds.), *The creative self*. Saint Louis: Elsevier Science.

Van Knippenberg, D., & Schippers, M. C. (2007). Work group diversity. *Annual Review of Psychology*, *58*, 515–541.

Vorauer, J. D., Gagnon, A., & Sasaki, S. J. (2009). Salient intergroup ideology and intergroup interaction. *Psychological Science*, *20*(7), 838–845. doi: 10.1111/j.1467-9280.2009.02369.x

Wharton, A. S. (1992). The social construction of gender and race in organizations. In P. Tolbert & S. B. Bacharach (Eds.), *Research in the sociology of organizations* (Vol. 10, pp. 55–84). Greenwich, CT: JAI Press.

Williams, K. Y., & O'Reilly, C. A. (1998). Demography and diversity in organizations: A review of 40 years of research. In B. Straw & R. Sutton (Eds.), *Research in organizational behavior* (Vol. 20, pp. 77–140). Greenwich, CT: JAI Press.

Zeidner, M. (1995). Personality trait correlates of intelligence. In D. Saklofske & M. Zeidner (Eds.), *International handbook of personality and intelligence*. New York: Plenum.

Index

A

B

C

D

E